D0445051

Lenin's Laureate

Transformations: Studies in the History of Science and Technology

Jed Z. Buchwald, general editor

Lenin's Laureate

Zhores Alferov's Life in Communist Science

Paul R. Josephson

The MIT Press
Cambridge, Massachusetts
London, England

For information about quantity discounts, email special_sales@mitpress.mit.edu.

Set in Stone Sans and Stone Serif by Graphic Composition, Inc. Printed and bound in the United States of America.

Library of Congress Cataloging-in-Publication Data

Josephson, Paul R.
Lenin's laureate : Zhores Alferov's life in communist science / Paul R. Josephson.
    p.   cm. — (Transformations : studies in the history of science and technology)
Includes bibliographical references and index.
ISBN 978-0-262-01458-8 (hardcover : alk. paper) 1. Alferov, Zh. I. 2. Physicists—Russia (Federation)—Biography. 3. Physics—Russia (Federation)—History. 4. Science and state—Russia (Federation)—History. I. Title.
QC16.A3417J67   2010
509.47'0904—dc22

                                                          2010005431

10 9 8 7 6 5 4 3 2 1

Contents

Introduction

When you look at the lights on your alarm clock or stop at a traffic signal, more than likely you see light-emitting diodes in action. LEDs consume 80 or 90 percent less energy than incandescent lights, and they last years longer. In addition to LEDs, in everyday life we encounter semiconductor lasers, solar cells, and transistors in a variety of communications and energy electronics—for example, CD players and mobile phones. These items reflect the practical applications of the heterojunction, a kind of semiconductor device that operates with very high efficiency at room temperature. The heterojunction was developed in the 1960s by a small number of physicists working in various countries. One of them, Zhores Ivanovich Alferov, shared the Nobel Prize in 2000 for his development of the heterojunction.

As of this writing, in 2010, the 80-year-old Alferov, based in St. Petersburg, remains active in research in the science of nanostructures and as a university educator, simultaneously serving in the Russian Parliament as a member of the Communist minority. Alferov believes that his role in the Parliament is to use his authority as a Nobel laureate to save Russian science and technology from further losses of funding, personnel, and research capacity—that is, losses beyond those that occurred immediately after the breakup of the Soviet Union. Alferov's life and career illustrate the rise of Soviet science and technology, the challenges that scientists endured in creating the largest research-and-development apparatus in the world, and the difficulties Russian scientists face today in securing funding for research.

By 1948, when he graduated from high school in Minsk with a gold medal, Alferov had demonstrated many of the personality traits that are essential to a career as a scientist. His teachers saw that he had a dedicated and disciplined mind, mastered difficult concepts easily, had immediate recall of relevant literature, and was capable of penetrating analysis. Alferov had not yet chosen a specific field of research, but he had a strong interest in electronics.

After World War II, under Joseph Stalin, the USSR engaged in a massive rebuilding program in which electrification was a central aspect. Alferov, like many other students, played a small part in that program, serving during his sophomore year in a brigade whose mission was to design and to build—largely by hand—the tiny Krasnoborsk hydroelectric power station in Leningrad province, which was to provide power for local industry. He then won a research prize that entitled him to visit any one of Stalin's "hero projects." He chose the Tsimliansk Dam on the Don River, whose huge size confirmed for him the glories of communist construction. These experiences convinced him to enter graduate school in the hope of making a personal contribution to the USSR's electrification programs.

After receiving a BS degree at the Leningrad Electrical Technical Institute, Alferov considered a wide range of first-rate institutes for his graduate work. Since 1917, when they had seized power, the Bolsheviks had underwritten the establishment of dozens of physics, chemistry, metallurgical, biological, and other institutes. The Cold War brought another burst of activity as new institutions, many of them connected with nuclear weapons research, were created. The nuclear enterprise attracted some of the best students; others were simply ordered to enlist in the closed, secret facilities. Alferov was determined to stick to electronics. He found a home and a career at the foremost center for research in solid state physics, the Leningrad Physical Technical Institute, which he entered at the beginning of 1953, shortly before Stalin's death. There, in the 1960s, he completed studies on heterostructures of gallium arsenide, later becoming director of the laboratory and then of the institute.

Many of the stories of the personalities of modern science focus exclusively on genius as the source of inspiration. What forces shape modern science? The qualities of the individual researcher (Alferov possessed mathematical aptitude, the ability to synthesize, and originality) enable a young scholar to embark on a career in physics, chemistry, biology, or mathematics. We commonly believe that the individual scientist, stimulated by colleagues or more generally by developments in a field of research yet also inspired by some internal logic or insight, is the prime engine of scientific discovery. Alferov surely shared many of these characteristics.

Yet by the early years of the twentieth century science had become "big science"—the science of research institutes, expensive apparatuses, and extensive government support. Such leading industrial companies as General Electric, DuPont, Westinghouse, and I.G. Farben had established laboratories in support of their business activities.[1] At universities, scientists, driven by the increasing costs of research and by difficulties they encountered in attacking the conundrums of modern science alone, successfully lobbied their governments and such newly established philanthropies as the Rockefeller Foundation for supplementary budgets, and began to work collaboratively in larger and larger teams.[2] Theoretical scientists were not immune from the need for team research using expensive equipment, but came to work closely with experimentalists who used such devices as particle accelerators and reactors to study subatomic processes and verify theoretical advances. In Alferov's field of research, these included devices for molecular beam epitaxy and metalorganic chemical vapor disposition developed in the 1960s and the 1970s.

The modern state played an important role in financing and shaping science. By the end of World War I, policy makers had recognized the need to support researchers to improve the increasingly industrial nature of military performance. By the end of World War II, with the big projects of rocketry, radar, and atomic weaponry, they recognized the necessity of assembling huge collectives of researchers in national laboratories, universities, and industrial facilities, and not only for national security but also for public health, economic stability, and even simply for the sake of science.[3] These national laboratories—Los Alamos, Argonne, Brookhaven, and Oak Ridge in the United States; the various Max Planck (formerly Kaiser Wilhelm) institutes in Germany; Arzamas-16, the Kurchatov Institute for Atomic Energy, and the Leningrad Physical Technical Institute in the former Soviet Union; and many others—remain icons of scientific excellence and state power.

Perhaps nowhere was government support so critical to the development of modern science as it was in the USSR. In 1917, on the eve of the Revolution, Russia was a kind of scientific colony of Europe. Its scientists frequently traveled to Germany and France to undertake research in modern laboratories and to complete their doctorates. The tsarist government had a narrow understanding of science's contribution to national well-being. Russian laboratories were poorly equipped for lack of support, and government bureaucrats often interfered in the hiring and firing of researchers on the basis of politics. The vast majority of the empire's citizens were illiterate, living on the edge of agricultural failure in a countryside far removed from urban, industrial lifestyles. A few philanthropies had been established, but they were relatively small and they focused most of their activities on areas other than science.

Upon seizing power in October of 1917, the Bolsheviks set out to create socialist institutions in the economy, in the educational system, in the arts, and in the sciences. Vladimir Lenin, Leon Trotsky, Nikolai Bukharin, and

Joseph Stalin recognized that they needed science and technology to build a modern industrial state. They were committed to expansion of the scientific enterprise. After all, they claimed that Marxism itself was scientific, and that modern science and technology were the foundations of communism. Over the course of Soviet history, this healthy interest in science generated the allegiance of Alferov and other students who later became leading figures of the scientific establishment.

Many scientists and other intellectuals, as members of the middle and upper classes, mistrusted or even hated the Bolsheviks (forerunners of the Communist Party), and millions of the most able citizens emigrated. A large number of scientists perished during the Revolution and the civil war. Yet those who remained welcomed the Bolsheviks' support for their institutions, if not their political interference in their activities. They reached an accommodation with the communist officials. They reestablished the contacts with Western scholars that had been disrupted by World War I, traveled abroad for study and conferences, and became full members of the international community.

How did someone like Zhores Ivanovich Alferov enter into and become a leading figure in the world's largest research-and-development enterprise? Alferov came from a modest family background. His father, Ivan Karpovich Alferov, was a man of working-class origins who became an official in the pulp and paper industry; his mother, Anna Vladimirovna (née Rosenbloom), was from a shtetl (a small town centered on small trade and agriculture) in the Pale of Settlement, a poverty-stricken, largely agricultural region.

This biography explores the rise of big science and technology in the twentieth century through the experiences of a leading scientist and against the backdrop of Soviet history from the Revolution to the USSR's collapse in 1991.[4] It discusses the personal, psychological, political, and economic forces that shaped science in the USSR, the upward mobility that was provided to several generations of aspiring scientists, the reverence in which Soviet leaders and citizens held science and scientists, the great human costs of the Stalinist development model, and the surprising inefficiencies of the centrally planned economy. It also seeks to explain the dire straits in which scientists find themselves in Russia today. Alferov lived and worked across this entire period, and we can understand his current activities as an educator and researcher, and as a member of the Russian Parliament, in light of this history.

On the eve of the downfall of communism, the Soviet Union had one-third of the world's engineers and one-fourth of its physicists. It tested atomic and hydrogen bombs years before Western observers predicted, and achieved parity with the United States in deployment of these weapons. It brought on line the world's first reactor, to power a civilian electricity grid (in Obninsk, southwest of Moscow, in 1954); it pioneered the tokamak (a fusion device); and in 1957 it beat the United States in the race to launch the first artificial satellite. Its scientists won ten Nobel Prizes in physics and one in chemistry, including three in the first five years of the twenty-first century. It built this scientific empire despite the human costs of Stalin's murderous industrialization and collectivization campaigns and the loss of 20 million citizens during World War II. Its scientists achieved impressive results in spite of the formation of a notoriously closed system that stifled educational freedom and creativity, emphasized military over civilian production, and placed collective achievement ahead of individual excellence. Many of the familiar faces of Soviet science (including the physicists Andrei Sakharov, Igor Tamm, Petr Kapitsa, Lev Landau, and Vitalii Ginzburg, all Nobelists), and scores of other less well known but supremely talented scholars, were arrested, lost their jobs, in some cases were imprisoned, and in others cases were executed for imagined crimes against the state.

Alferov watched in horror as the USSR fell apart after political and economic crises, then as the Russian government under Presidents Boris Yeltsin and Vladimir Putin failed to support science to the level required to preserve his country's leading position. He believes that incompetent bureaucrats maliciously seek to dismantle what remains of the Soviet scientific heritage, including his beloved Academy of Sciences, his home for research on solid state physics for more than 50 years. In this biography, which is based on unique access to archives, internal documents, and a series of interviews with Alferov and other individuals who have been connected with his life and career, I try to give the reader a sense both of Alferov's life and of Soviet and Russian science generally, hoping to contribute to our understanding of big science, politics, and the modern state before, during, and after the Cold War.

The Soviet intelligentsia consisted of individuals who self-consciously recognized as their central function to develop, protect, and spread culture. It included, as well as members of the scientific and engineering communities, creative artists—writers, poets, painters, composers, and others. Both the scientists and the artists read and wrote voraciously. They set an example for Soviet citizens, who also learned to read voraciously; on buses, trams, and subway cars, nine out of ten individuals opened a book to read as soon as they took a seat. Through literature and poetry, writers often addressed issues of politics and society that were taboo in other forms of academic writing, even during the Stalin era of strict control over publications of all sorts. They triggered debates about the nature of the Soviet political system, about the centrally planned economy, and even about the morality of the "new" Soviet man and woman. Zhores Alferov, too, was an avid reader from childhood. He enjoyed poetry, short stories, and novels that enabled him to consider more fully the predominant Soviet mores, and to ponder veiled and not-so-veiled criticisms of Soviet society. We will therefore have the opportunity to consider the works of a few of Alferov's favorite writers and poets (Vladimir Mayakovsky, Mikhail Zoshchenko, Ilya Ehrenburg, Vassilii Grossman, and several others), the importance of these works for Alferov, and the significance of the works for the Soviet social and political order.

Alferov—a Nobel laureate, a founder of modern solid state physics, director of an institute known as the cradle of Soviet physics (the Leningrad Physical Technical Institute, abbreviated LFTI), one of eleven vice presidents of the Soviet (now Russian) Academy of Sciences, a central figure of the scientific establishment during the Gorbachev years, and a member of the Soviet and then the Russian parliament (first as a representative of the Academy of Sciences and now as a deputy of the Communist fraction)—a man whose devoted Bolshevik parents grew up not far from the painter Marc Chagall's home town of Vitebsk—came into scientific adulthood under Stalin, made his first scientific breakthroughs during the era of Khrushchev while still a junior researcher, and rose to prominence during the Brezhnev era. His life story reveals the strengths and weaknesses of Soviet science and technology, the vagaries of doing research in relative isolation from the West, and yet the deep respect for science among Soviet citizens and leaders that created an environment for path-breaking discovery. One example will illustrate: In spite of Alferov's pioneering efforts to create heterojunctions, the Soviet Union failed to provide adequate support for efforts to seek applications for these devices. But heterojunctions turned out to have applications in semiconductor transistors, optoelectronic instruments, and other devices, and in many cases these applications in lasers, communications, mobile phones, CDs, DVDs, and solar power first occurred outside the USSR.

Alferov began his career in an environment in which advanced research in nuclear, theoretical, solid state, and low-temperature physics was accompanied by ideological attacks and political interference in science. In the 1930s, Stalin introduced major changes in the policies of the government toward science. On the one hand, as part of the industrialization drive, the government substantially increased funding for research and development, especially in those institutes (including the LFTI) that were now subjugated to industrial ministries and harnessed to overly ambitious five-year plans. This enabled scientists to embark on research in such new fields as nuclear and low-temperature physics and in other fields that seemed to bureaucrats to have few short-term applications for industry. At the same time, the Bolsheviks grew increasingly fearful of ubiquitous alleged enemies of the state, including "tsarist remnants" in science and education. They set out to infiltrate and take over research institutions, carried out ideological campaigns to cleanse the institutions of "hostile elements" and "class enemies," and created a top-heavy bureaucracy to administer science—a bureaucracy that stifled the creative impulse of many a scientist. A kind of scientific autarky resulted. Many of Alferov's teachers thus commenced their careers in a guarded atmosphere in which they were challenged to demonstrate their allegiance publicly at every turn.

Scientists anticipated that their contributions to victory over Germany in World War II would contribute to a relaxation of control over their research programs at home and to more contact with colleagues abroad. But the defeat of Germany provided no breathing space. In the postwar years, Stalinist ideologues redoubled their campaign against idealism and against "kowtowing" before the West. Biology—which fell under the spell of the quack agronomist Trofim Lysenko, and which saw genetics purged from high schools and universities—was not the only science to suffer the consequences. In physics, enemies of relativity theory and of quantum mechanics sought to purge practitioners of modern science from Soviet research institutes, and many people argue that only the USSR's quick success in building an atomic bomb saved physics from a wholesale purge. Physicists at the LFTI and at similar institutes were not immune. Alferov avoided political and philosophical controversies, assiduously focusing on his research. He has always considered himself apolitical, even though he has served Communist Party organizations at the institute, provincial, and national levels.

Though in his published autobiography (*Nauka i Obshchestvo*) he downplays his scientific talents and the significance of his entry into research, Alferov has been at the center of Soviet physics ever since he completed high school. He joined the Leningrad Electrotechnical Institute (LETI) in 1947 without taking an entrance exam, having been awarded a gold medal for finishing at the top of his high school class in Minsk. The LETI had a glorious pre-revolutionary history, but Alferov dreamed of moving to the Ioffe Institute because of the reputations of its director, Abram Ioffe, and such leading personnel as the theoretician Iakov Frenkel and the experimentalist

Vladimir Tuchkevich. Founded in 1918, this institute employed about 40 members and 25 corresponding members of the elite Soviet Academy of Sciences, scores of recipients of Stalin, Lenin, and State Prizes, three Nobel laureates (Petr Kapitsa, Lev Landau, and Nikolai Semenov), the head of the Soviet atomic bomb project (Igor Kurchatov), a central figure in the design of the Chernobyl-type reactor and a president of the Academy of Sciences (Anatolii Aleksandrov), and the creator of the "big bang" theory of the origin of the universe (George Gamow). Under the leadership of its personnel, another dozen or so institutes were established. The Ioffe Institute's specialty was solid state physics, but its scientists also contributed to the development of quantum mechanics, nuclear physics, and a number of other fields.

Alferov's initial work at the LFTI, done in Tuchkevich's laboratory, involved the physics of powerful semiconductor thyristor valves. Because of his successful completion of this project, Aleksandrov asked that Alferov be sent to Severodvinsk, a submarine- and ship-building city on the White Sea, to consult in the creation of rectifiers based on germanium crystals for the first Soviet nuclear submarine, the *Leninskii Komsomol*. In addition, he consulted with the LFTI's new director, Boris Konstantinov, on lithium isotope separation for the hydrogen bomb project. Clearly, Soviet researchers were centrally involved in military research and development during the Cold War. But Alferov's main field of interest was the quantum physics of semiconductor instruments. Only briefly connected with research on nuclear topics, he consciously chose not to enter departments connected with nuclear research. Alferov was interested in non-classified research. He sought open and direct contacts with foreign scholars and expansion of publication in open journals. Throughout his career he worked on a number of scientific journals, several of which were based at the LFTI. He served as editor of the *Journal of Technical Physics (Letters)* (*Pis'ma v Zhurnal Tekhnicheskoi Fiziki*) and of *Physics of the Solid State* (*Fizika tverdogo tela*), and on the editorial boards of several other journals. Alferov believed that the overriding secrecy of scientific research in the USSR and the systematic interference that prevented the publication of journals in a more timely fashion and generally stultified scientific progress.

In the Khrushchev era, the conflict between reformism and conservatism that characterized broader Soviet political trends directly affected the work of Alferov and his colleagues. In 1950, at the peak of Stalin's power, Abram Ioffe was ignominiously forced to retire from the institute he had founded and had led over the course of three decades. Anti-Semites and nationalists working within the scientific apparatus of the Academy and the Central Committee of the Communist Party had tired of his authority, and he had come increasingly under attack for his defense of modern physics. He was replaced by Anton Komar. Though Komar may have been an able scientist and administrator, he was not Ioffe's equal, and he was better known for his allegiance to conservative Soviet principles. Many scientists considered him an outsider and found fault with his style of leadership. However, Komar ably extended the LFTI's programs into a search for qualitatively new semiconductor electronics, the development of technologies for the space program, and the working out of methods for separation of light isotopes for hydrogen bombs. Boris Konstantinov replaced Komar as director of the LFTI in 1957. By all accounts, Konstantinov was an excellent and beloved scientific leader who presided over institutional growth that benefited Alferov, in part through the creation of his own laboratory.

Khrushchev promoted reforms in a wide-ranging but inconsistent program of de-Stalinization. Some of the reforms affected science. No longer fearing scientific autonomy and independence of thought, and recognizing the importance of specialists in a campaign to improve industrial production through automation and mechanization, the government promoted a fourfold increase in the number of scientists in the 1950s alone. A number of research centers were founded, including the Siberian city of science, Akademgorodok.[5] The government approved the removal of narrowly technical and engineering specialties from the Academy and their affiliation with industrial ministries. And the Communist Party permitted more scientists to travel abroad more frequently and to exchange preprints liberally, especially after the 1955 Geneva Conference on the Peaceful Uses of Atomic Energy. Over the next 30 years, Alferov, increasingly an internationalist in his own views, attended conferences in Europe and took fellowships in the United States that enabled him to publicize Soviet priority in discovery.

In 1961, Alferov defended his Candidate of Science degree, which focused on germanium and silicon rectifiers. He then turned to heterostructures, beating scientists at Bell Laboratories, IBM, RCA, and other industrial and university laboratories to discoveries in that field. International contacts and competition enabled him to ensure the priority of Soviet research. "My most productive years were the 1960s," Alferov told me, "not only because I was a thirty-to-forty-year old, but there were exchanges of scientists. We were open."

Alferov's activities in Party organizations dated back to his youth. With an excellent memory, he was able to remember speeches, slogans, and programs with ease. His faith in the regime remained constant. His certainty that communism supported science and technology for the benefit of the masses grew in the course of his career. He gladly joined the Komsomol (Communist Union of Youth) as a young boy during World War II, but to the discomfort of his father—a Bolshevik since the fall of 1917—he put off joining the Communist Party until 1967. He says he had no desire to join the Party, and adds that he did not consider it important or necessary to his career, although he respected those at the LFTI who belonged to the institute's Party organization. He believed that, for the most part, the Party organization was dedicated to supporting research. Especially after the Stalinist legacy had worn off, the organization had lost some its paranoid vigilance. His divorce from his first wife had been messy, and he expected to encounter at least some discussion of it when he finally joined the Party, but no one asked him about this episode; he claims that Director Tuchkevich and the others were concerned about his contribution to the organization, not about any gossip.[6] In any case, Party membership was crucial to career advancement in any area of the economy. Science was no different. By the 1960s, roughly 70 percent of institute directors belonged to the Party. (In 1929, on the eve of Stalin's industrialization and collectivization campaigns, only a few of the USSR's leading scientists were Party members.) And by the mid 1970s, the Soviet government and the Party apparatus had become nearly technocratic.

Party membership was not indispensable to success in science, but it did not hurt one's career in the Brezhnev era, an era of limited political and social reforms. The Brezhnev regime had conservative domestic policies, including attacks on dissidents and restrictions on human rights. The military assumed a central role in Soviet society, not only in pursuit of parity with the United States but also in educating society about the need for vigilance and devotion to the state. Dissidents sought to force the government to uphold the provisions of the Brezhnev Constitution, which had guaranteed various civil liberties. But the Party would have none of this, and it pursued dissenters in the artistic, literary, and scientific communities with the full power of the KGB. At a number of research institutes, including the LFTI, scientists engaged in dissident activities. One such scientific dissident, the physicist Andrei Sakharov, was isolated and eventually exiled. Like J. Robert Oppenheimer, who was ostracized by the American military establishment for his opposition to the hydrogen bomb, Sakharov and other dissenters fell under constant surveillance and were isolated.

Alferov joined the Soviet Academy as a full member in 1979. Though he was not a dissident by any stretch of the imagination, his colleagues considered him among the reformers of the Academy. Later, in the Gorbachev

era, he sought, with other reformers, to increase the dynamism of the research enterprise and the flexibility of the scientific bureaucracy. Yet, while
the openness of the Gorbachev period invigorated literature, the arts, and
politics and contributed to discussion of the role of science in modern society and the costs of the Soviet model of economic development, Gorbachev's perestroika (restructuring) had, on the whole, a negative effect on
science. This was largely for financial reasons: the recession and the sky-
rocketing inflation that followed the failed reform efforts left institutes and
researchers without adequate support for equipment, reagents, or salaries.

In 1989, in the first open elections to the Congress of People's Deputies,
Zhores Alferov, Andrei Sakharov, and Vitalii Ginzburg were elected. (Sakharov had been allowed to return to Moscow from internal exile in Gorky.)
These men debated the nature of the Soviet system, what role the Communist Party ought to have in the empire's future, and how to improve the
USSR's scientific performance, which had begun to wilt under the weight of
a stifling bureaucracy that slowed publication and innovation.

At first Alferov reveled in the excitement of the Gorbachev reforms. He
spoke at a variety of forums, and he published a series of articles on the
meaning of perestroika for the rejuvenation of science. In 1990 I had lunch
with him in the Academy of Sciences' dining hall on Lenin Prospect in
Moscow. He shared his excitement about the ongoing internationalization
of science and suggested that a real flowering of Soviet science would follow. Soon, however, as the government's attention to and support for science dwindled, he felt a need to defend science in the Congress of People's
Deputies and to prod the government to maintain its support. While a
member of the parliamentary committee on science and education, Alferov
was continually frustrated in his efforts to stem the decline of science.

In December of 1991, a few months after a failed coup, Mikhail Gorbachev determined that he was powerless to hold the Soviet Union together and resigned as its president. The Soviet empire disappeared nearly
bloodlessly, and Boris Yeltsin, now the president of Russia, faced burgeoning economic and political crises that told upon the scientific enterprise.
Yeltsin's policies were inconsistent, and his relations with the parliament
were poor. The economy entered a deep recession, and well-placed individuals stripped its assets wantonly. Inflation meant there was little money for
salaries or utilities, let alone for research. Alferov struggled, as director of
the LFTI and as a parliamentary deputy, to gain the Russian government's
support for science. The failure of the government to fund science adequately led Alferov to enter the Communist fraction; to him it seemed to
be the only political party willing to defend science and higher education.

At this writing, Alferov continues his energetic pursuit of scientific excellence under conditions that would have encouraged many people of such an age to retire. Funding for research and development is perhaps one-sixth what it was when the USSR disappeared in 1991. According to senior scientists in the Russian Academy of Sciences, neither President Boris Yeltsin nor President Vladimir Putin concerned himself with science to the extent that was necessary, seeing it only for its immediate utility and desiring to control the Academy's real estate and other assets despite the pleadings of Alferov and others. President Dmitri Medvedev's policies remain unclear. Many talented researchers have gone abroad. The aging of the scientific community and the pursuit of business careers by young people have led to a growing generation gap. Funding for foreign travel to attend conferences has dried up. Not even the support of Western governments, foundations, and firms—which Alferov welcomed—has protected the scientific establishment.

Russia inherited the Soviet Academy of Sciences, the Soviet Union's three major universities (Moscow, Leningrad, and Novosibirsk), and scores of lesser-known scientific and educational institutions. In the 1990s, Russia drew the attention of Western grant and contract programs that were established to prevent the migration of scientists with weapons knowledge to so-called rogue nations. But many laboratories indeed lost the ability to compete with those in the West. Modern equipment, reagents, and even journal subscriptions have become a luxury in many institutes, although Russia's economy has grown rapidly and the government should have adequate resources to support the scientific enterprise.

From his positions as vice president of the Academy of Sciences, head of the St. Petersburg Scientific center, and a member of the Parliament, Alferov succeeded in preserving the position of the Leningrad Physical Technical Institute, in expanding the role of a "lyceum" for promising high school students, and in creating a new university to ensure a steady flow of talented young people into the sciences at his Scientific Education Center (Nauchno-Obrazovatel'nyi Tsentr, abbreviated NOTs)—a new facility, across the street from the LFTI, that combines education and research as American and European universities do. Alferov built NOTs with government support and with funds from his Nobel Prize and from the Siemens Corporation. He sees protecting science as his mission. For him science is a national treasure, and he believes that only the communists are truly dedicated to protecting it. He accuses government administrators of incompetence if not treason for the failure to support science. But his defense of communism has drawn criticism within Russia and from elsewhere. Vitalii Ginzburg accuses

Alferov of forgetting the great human costs of collectivization, the purges, and the labor camps, and calls communists the major danger to Russia and the entire world. Yet Alferov remains a committed communist and sees the achievements of the Soviet empire—a society of dynamic social mobility, with a system of universal education that eradicated illiteracy in only a few years, and with modern health, educational, and scientific institutions—as evidence that the communists understood what was crucial to support the country's future, whereas the Yeltsin and Putin governments have not addressed the problem of how to bring Russia into the twenty-first century. Alferov argues that Russia's future will be based on the achievements of science and technology, and specifically nanotechnologies, not on selling the country's vast natural resources of oil, gas, and timber. In part because of his determined effort to convince government bureaucrats of the errors of their science policy, he has become increasingly estranged from the political establishment.

The heterojunctions that Alferov developed the early 1950s and the 1960s as a junior scientist at the LFTI led to advances in semiconductor lasers, diodes, and other devices. The miniaturization that heterojunctions made possible prepared the way for today's nanotechnologies. As Alferov wrote (in English) in a review article, "The most fruitful idea for development of physics and applications of heterojunctions for different electronic components was the concept of double heterostructure proposed by us at the beginning of 1963. Electron and optical confinement, super injection phenomena became new tools to control electron and light fluxes in crystals. Many new optoelectronic and electronic devices were created on the base of heterostructures. Of utmost importance was the making of room temperature CW [continuous operation] lasers, efficient light emitting diodes and high-speed transistors."[7] Alferov has similar dreams for future economic growth based on nanostructures.

1 Childhood

Revolution, Civil War, and the Rise of Soviet Science

A wail spreads over the village. The cavalry is trampling the grain and trading in horses. The cavalrymen are exchanging their worn-out nags for the peasants' work-horses. One can't argue with what the cavalrymen are doing—without horses there can be no army.

—Isaac Babel, *The Red Cavalry Stories*

If there could be a typical family in revolutionary Russia—a time of radically shifting borders, occupations, and political allegiances, of war and civil war, pogroms, epidemics, starvation, and of violent efforts to create a new, socialist, industrial society on the foundation of an agrarian, peasant economy and a failed autocracy—then Zhores Alferov came from a typical family. His father, Ivan Karpovich Alferov—a soldier and a border guard for the Cheka (the forerunner of the KGB, founded under Lenin), and a convinced Bolshevik—took advantage of the opportunities for social mobility offered by the revolution to become a leading official in the pulp and paper industry. His mother, Anna Vladimirovna (née Rosenbloom), came from the Pale of Settlement, to which Jews had been confined by the tsars beginning with Catherine the Great. Having married in the mid 1920s, the young couple moved from one pulp, lumber, and paper mill to another, living at one time or another on the Dvina River, the Ob River, and the Volga River, in Arkhangelsk province, in Siberia, in Leningrad, in Stalingrad, and eventually, after World War II, in bombed-out Minsk, the capital of Belarus (formerly called, in English, Byelorussia or Belorussia).

Because of his competence, his dedication to work, and his devotion to Bolshevism, Ivan Karpovich Alferov always found himself with connections to the Party elite, although he did not consider himself a member of that elite. His connections helped him to find good jobs and decent homes

Zhores Alferov's maternal grandfather, Vulf (Vladimir) Rosenbloom, in Kraisk, in the Pale of Settlement, 1910.

in a time when many Soviet citizens lived on the edge. The cultural revolution that transformed institutions of science and higher education under Stalin enabled him to move through the ranks to assume a leading managerial position, and in the future it would provide opportunities for his son Zhores to gain entry to one of the world's foremost centers of research in solid state physics.

Ivan Karpovich Alferov was born September 20, 1884, in the little town of Chashnika in Vitebsk province, in present-day Belarus, into the family of a landless lower-middle-class merchant, Karp Pavlovich Alfer, and his wife, Maria Grigorievna.[1] In the winter Karp worked as a bootmaker, in the summer as a raftsman. When Ivan enlisted in the tsarist army in January of 1915, the official filing out his papers changed the surname to Alferov to

Mariia Grigorievna Alfer, Zhores Alferov's paternal grandmother, 1938.

make it sound more Russian and less German. Ivan Karpovich Alferov had two brothers, Valerian and Volf, and a sister, Tatiana.

According to his Uncle Valerian, Ivan was a bright and lively little boy. He sang in the church choir. A local teacher who had noticed his intelligence arranged for his free education in a municipal school. He finished three of the four grades, and thus was quite educated for his time and his social class; in 1897 the average literacy rate in European Russia was 30 percent, among the rural population 26 percent, among men 43 percent, and among women 22 percent.[2] When his father worked as a lumberjack, Ivan joined him on the spring lumber float on the western Dvina River from Vitebsk to Riga, learning to play cards and to speak Latvian. At age 14, Ivan started work in the Chashnitsk Paper Factory. He would be connected with the pulp and paper industry for his entire adult life.

By the end of the nineteenth century, Russia had entered the Industrial Revolution. Iron and textile mills, mines, and factories grew in number and size, especially around St. Petersburg and Moscow, and also around Baku, in Azerbaijan, where rich oil fields had been tapped. Sergei Witte, Tsar Nicholas II's minister of finance, saw his dreams of a modern Russia realized in the construction of the Trans-Siberian Railroad. He knew how railroads had contributed to the opening of the American West, and he believed that the Trans-Siberian project would attract European and Russian investors, and not only to the railway system. Even before the formation of the Bolshevik Party in 1903, the future leader of the Bolsheviks and the USSR, Lenin, had published a book (*The Development of Capitalism in Russia*, 1895) arguing that Russia had become an industrial nation and that the time for revolution was near.[3]

Yet in most regions and most sectors of the economy Russian industries lagged far behind those in Western Europe and the United States. Most enterprises tended to be smaller than their European counterparts, and industry faced the challenge of attracting skilled laborers. As has already been noted here, only 40 percent of males were literate, and most workers were at most one generation removed from the Russian countryside. In addition, aside from the railroad the tsarist government had difficulty deciding which projects to support to "jump start" the economy, and because of the vagaries of tsarist policies the government had great difficulty in attracting foreign investors.[4]

Political unrest contributed to economic uncertainties. Tsar Nicholas had no understanding of the concerns of the people. He considered the nation his patrimony. Letters he wrote to his wife Alexandra during periods of crisis, including the failed revolution of 1905 and World War I, address questions of family life, the weather, his hobbies, or what he was reading, not the challenges facing his rule. In 1904, Nicholas and his advisers sought war with the Japanese over influence in Manchuria. They determined to transform growing public dissatisfaction with Japan into nationalism through what they believed would be a "quick little war." But the Russo-Japanese War turned into a fiasco. The Japanese military was more than a match for the tsar's armies and navies. Strikes spread through St. Petersburg, Moscow, Ivanovo-Voznesensk, and other major industrial centers. Facing certain revolution, the tsar capitulated, agreeing to the formation of a duma (parliament) with limited powers. Four parliaments, each increasingly conservative, served until 1917, when the tsar abdicated; Nicholas limited the franchise to more and more elite segments of the population to establish a duma to his liking, and in an effort to subvert the public will.[5] Russia would not see a true parliament again until 1991.

Peasants and other laborers continued to migrate to major cities in search of work. In 1910, Valerian Alferov went to Petersburg, where he joined the Siemens company. After the Russian Revolution, the former Siemens factory became Elektrosila, the Soviet Union's leading manufacturer of turbogenerators and other electrical equipment. Valerian worked for that company his entire life as a senior metal craftsman and fitter—even during the blockade of Leningrad during World War II, when Elektrosila workers had to toil in bombed-out buildings. In 1912 Ivan followed him to St. Petersburg, where he found work in an envelope factory and then as a longshoreman.[6] Like tens of thousands of other workers, Valerian, Ivan, and several friends crowded into rented room that became a site of revolutionary activity.

At the outbreak of World War I, many Russian citizens were initially filled with patriotic feelings against Germany. They backed the tsar and Russia, and assumed that the war would be quick and victorious. But the war tipped the balance against the regime. Russia lacked rolling stock to get troops and supplies to the front. It lacked competent non-commissioned officers. Many of the soldiers were illiterate peasants who believed they ought to be working in the fields. Exporting its great natural resources, Russia had relied heavily on Germany industry for finished goods, including

The brothers Ivan (first from right) and Valerian (second from right) Alferov with friends in St. Petersburg, 1913.

explosives. Now it had to develop indigenous munitions capacity rapidly. Soon the war effort bogged down, soldiers began to desert, and Russia's cities were choked off from supplies of food and fuel. Because of the tsar's failure to achieve any significant political and economic reform, his miserable conduct of the war with Germany, and the growing influence of the monk Gregorii Rasputin over his wife and his advisors, even increasingly conservative parliamentarians lost all hope in him. This led one moderate deputy (Pavel Miliukov, founder of the Constitutional Democrats party) to ask, in November of 1916, "Is this incompetence or is this treason?"

In the first days of the war, Ivan joined the army voluntarily. He returned to Chashnika to sign up, having been told that he must do so in the place of his birth. He was sent to the cavalry. Zhores recalls that his father served in the Fourth Hussars of the Imperial Army's Mariopol Regiment. He fought almost the entire war at the northwest front, and longest of all in Latvia. He was by all accounts a courageous solder, winning a Gregory's Cross and rising to lieutenant. Because he was literate, he was sent to ensign's school, but he missed the action and returned to his regiment at the front. In February of 1917, spontaneous rebellions over food and fuel shortages spread throughout the nation. In St. Petersburg, the police and the Cossack regiments that had been sent to protect tsardom quickly went over to the side of women, children, and workers, and the tsar was informed that he had no choice but to abdicate to a provisional government.

After the February revolution, Ivan opposed the provisional government for insisting upon following tsarist treaty obligations.[7] In this case the obligations meant continuation of a costly, unpopular, and unsuccessful war. Only the Bolsheviks among the many parties were strictly antiwar, a factor that led Ivan to join them and not the Mensheviks, the centrist Cadets (Constitutional Democrats), or some other party. The provisional government sought legitimacy at all costs, but in actuality shared power with the Petrograd Soviet of Workers' and Peasants' Deputies. (At the outbreak of World War I, the government had changed the name of the capital from "St. Petersburg" to "Petrograd" on the notion that the older name sounded too Germanic; in 1924, on the death of Lenin, the city became "Leningrad," and in June 1991 the residents of the city voted to return the name to "St. Petersburg.") This led to a situation of dual power, the government having authority but no power and the Soviet having power but no authority. On top of this, the officials of the provisional government were so consumed by issues of legality that they failed to take decisive action against representatives of counterrevolution on the right or representatives of revolution on the left, preferring to wait for national elections to some representative body (the Constituent Assembly), and their authority rapidly waned.[8] In the countryside, anarchy rapidly spread as peasant soldiers returned home; the peasants confiscated land that they believed belonged to them, burned the landlords out, and killed many of them. Ivan Karpovich Alferov's opposition to the war led to his arrest in July of 1917. He was sent to the Dvina Fortress prison for a month and a half.[9] During these so-called July Days, Petrograd workers and soldiers rioted against the provisional government. After hesitating, the Bolsheviks ordered armed insurrection, thinking the time for revolution had come. But government police put down the revolt and ordered the arrest of the revolutionaries. Trotsky, Lenin, and other Bolsheviks went underground until September of 1917, when the provisional government released those in prison to fight a feared counterrevolution from the right under General Alexander Kornilov and his allies, who marched unsuccessfully on Petrograd.[10]

Freed from prison, Ivan Karpovich Alferov joined the Bolsheviks, a decision he stood by even when he faced arrest during the Stalinist purges of the 1930s. He joined the Military Revolutionary Committee, the Bolshevik organ of insurrection and the predecessor to the Red Army, in late September of 1917, serving under E. M. Sklianskii. Later, during the civil war, Sklianskii became deputy chairman of the Military Revolution Committee in the Commissariat of War under Trotsky. During the Stalin era, Trotsky was written out of Soviet history books, and even excised from published photos. Yet Trotsky was the military savior of the revolution. Ivan Alferov himself met with Trotsky several times, and frequently reminded his son of Trotsky's contribution to Soviet power through the creation of the Red Army. Zhores Alferov told me: "My parents did not like Stalin at all. My father especially respected Trotsky and knew his crucial role in the Red Army, enlisting tsarist officers into it and saving the revolution."[11] Ivan Alferov personally heard Lenin and Trotsky speak. His dedicated service led his regiment to elect him as a delegate to the Second Congress of the Soviets in Petrograd in October, at which Lenin submitted decrees on peace and land and formed a new Soviet government.

Lenin and Trotsky determined that the provisional government was vulnerable, and ordered troops loyal to the Bolsheviks and under control of the Soviets to seize power in October. Relatively bloodlessly, they took Petrograd; then, with more blood, they took Moscow, declared the socialist revolution complete, and set out to establish a socialist order. Over the next few months, in the name of the new Soviet government, Lenin issued a series of proclamations and decrees on peace, on the nationalization of property, on the right of national self-determination, and on the right of the peasants to own the land they farmed (endorsing what, de facto, had happened in many regions).

Having seized power in a coup, the Bolsheviks faced opposition from all sides and one crisis after another. In January of 1918, when elections to a national Constituent Assembly gave the Socialist Revolutionaries (a party that appealed to the peasantry) a majority, the Bolsheviks shut down the assembly after its first meeting. Over the next few years, the Bolsheviks moved to exclude leftist parties from political life, arrested many leading Socialist Revolutionaries, Mensheviks, and others, and declared the parties illegal. On the right, monarchists and nationalists of many stripes banded together as "Whites" to bring the Bolsheviks down. They saw the Bolsheviks as interlopers or worse. The Bolsheviks also infuriated the Germans, Trotsky having famously refused to negotiate a peace treaty with the slogan "No war, no peace." The Kaiser's armies attacked and quickly overran much of Ukraine and other western regions of the nation, and the Bolsheviks had to sue for peace in the Treaty of Brest-Litovsk (1918), losing considerable territory. (The treaty signed at the end of World War I required Germany to give these gains back to Russia.) The ensuing civil war between Reds and Whites led to millions of deaths by war, disease, and starvation, to mass dislocation, and to flight from the cities. The American Relief Administration had to step in to save millions of people from famine.[12] In spite of being outnumbered, the Bolsheviks prevailed by 1920 owing to their unwavering and competent political and military leadership under Lenin and Trotsky, the splintered nature of the opposition especially among the Whites, and their control from the start of the central industrial heartland, with its industry and railroads.

Ivan Karpovich Alferov was not only a courageous soldier in World War I, and not only a Bolshevik, but also a hero of the civil war, one of many eventually elevated to that stature. Having joined the Red Army in 1918, he served until 1923. In 1918, after informing his superiors that his cavalry group was ready to engage Whites, monarchists, and other enemies of the Bolsheviks, he was sent to the Donbass (Don River Basin), a coal-rich area that encompasses the southern border region of Russia and Ukraine, to combat one of the first of the many uprisings and skirmishes against the Bolsheviks that eventually led to a civil war. There Alferov and his comrades engaged General A. M. Kaledin, a Russian aristocrat who had served as a general in the tsarist army until dismissed in May of 1917. In July of 1917 the Don Cossacks had elected Kaledin their leader. In January of 1918, Kaledin joined forces with the monarchist generals Alekseev and Kornilov. Many of the Don Cossacks believed that Kaledin had betrayed their cause to the White armies. The Cossacks had traditionally opposed the monarchists, since the tsars had always attempted to subjugate them to the imperial army. Kaledin, disturbed by the loss of the autocracy and feeling rudderless in the growing tumult, killed himself in February of 1918.[13]

In a battle with the Whites, probably at the beginning of 1918, Ivan and his comrades were taken prisoner. As Ivan told Zhores, the Cossacks forced the prisoners to strip to their underwear and locked them in a shed. In morning they would have held a perfunctory court hearing, which would have resulted in a sentence to death by firing squad. A number of men froze to death that night, but Ivan and a comrade crept off into the night as the guards slept. Barefoot and nearly naked, they entered an empty house, stole felt boots, fur hats, and winter coats, and made their way back to their squadron.

Ivan Karpovich Alferov witnessed firsthand the anarchy and human suffering of the civil war. After returning from the Don to his home in Chashnika, he was soon called to Vitebsk by the deputy chairman of the regional Party executive committee, I. M. Vareikis, who informed him that, as a horseman who could speak Latvian, he was to form a Latvian battalion. (Vareikis, whom Ivan considered a friend, worked in agitation and propaganda, then in the secretariat of Central Black Earth region, and by 1936 was transferred to the Primorskii [Far East] region. In 1937 or 1938 he was accused of being an "enemy of the people" after expressing doubts about the arrests of the Great Terror; he then disappeared.) In the summer of 1918, Alferov's battalion engaged not only Whites but also Socialist Revolutionaries and other parties of the left. Fighting next in the Caucasus, he rose to commander of a cavalry regiment. At the end of the civil war, he was in Azerbaijan commanding a cavalry regiment of the Eleventh Army. He was wounded twice, and his horse was hit by shrapnel and fell on him.

After demobilization from the Red Army, Ivan served in the Cheka as a border guard. His son told me that these were not the same Chekists who meted out summary judgments against perceived enemies of the state during Lenin's notorious Red Terror, yet Ivan's service involved, as he wrote in his personnel folder, "struggle with smugglers,"[14] and probably included violent confrontations. During the civil war, the Bolsheviks instituted War Communism, under which they used military means to control the "commanding heights" of the economy and summarily punished workers for absenteeism and tardiness. During the Red Terror, as an adjunct to War Communism, the Bolsheviks executed thousands of perceived enemies of their rule.[15] According to Zhores Alferov, his father avoided these actions, although as a Cheka border guard he must have found it necessary to use his weapons. Early in the spring of 1922, Ivan was sent to the vicinity of the little town of Kraisk, about 80 kilometers northwest of Minsk, on the Soviet-Polish border, in the former Pale of Settlement. Seeking living accommodations, he wandered to the town's outskirts. He happened upon a house that impressed him with its simplicity and cleanliness, in spite of its dirt floor. A young woman who eighteen months later would become his wife lived there.

Zhores Alferov was an avid reader by the time he was five or six. He sought out literature that captured Soviet life and its heroes. As we shall see, he loved the classics of Russian and Western literature, loved such idiosyncratic representatives of Soviet poetry and prose as the poet Vladimir Mayakovsky and the satirist Mikhail Zoshchenko, and even became enamored of Socialist Realism, a new style, promoted under Stalin, in which the authors—pressured by the state authorities—left no doubt about the positive heroes of socialism and their predatory enemies. He sought the latter genre for its reflection of the lives of his parents. In one of the pioneering novels of the Stalinist genre of Socialist Realism, *How the Steel Was Tempered* (1930), Nikolai Ostrovsky describes the experiences of a young boy, Pavel Korchagin, who could be a model for Ivan Karpovich Alferov. Korchagin, a boy of modest working-class background, comes of age during the war, the revolution, and the civil war in action that takes place around the Ukrainian border town of Shepetkova in the region of the Pale. Zhores told me that he enjoyed this book as a young boy, perhaps for its resemblance to his father's career path.

The Pale of Settlement

Anna Vladimirovna Rosenbloom was born April 10, 1900, in Kraisk, Vilensk province, a town of almost 700 people (according to the 1897 census) in the Pale of Settlement.[16] The Pale, as has already been mentioned, was a poverty-stricken, largely agricultural region. More than 5 million of Europe's nearly 9 million Jews lived in the Russian empire, 90 percent of them in the Pale, which had hundreds of shtetls. The Pale included the provinces of Kovno, Vilna, Mogilev, Minsk, and Kielce, and extended southward all the way to Crimea. Within the Pale, Jews had to have a special permit in order to live in a major city (Kiev, Nikolaev, Yalta, or Sevastopol), and had to have a special passport in order to travel outside the Pale. Anti-Semites carried out murderous pogroms in the Pale, especially in the periods 1881–1883 and 1903–1906. To escape poverty, many young men left the Pale to seek employment in cities, especially Petersburg,[17] and millions would eventually emigrate to America and elsewhere.

Anna's father, Vladimir, known as Vulf, did odd jobs and made deliveries by wagon. Her mother, Khaia, was a religious and hard-working woman. As Zhores recalls being told, on one occasion a wealthy local Polish land-owner rewarded Vulf with a gift of three Karelian birches for always being on call and working hard. (Karelian birch is considered one of the finest and most valuable woods of the Russian north.) Vulf cut them down and milled them into boards and planks, which he then sold. He then bought a proper house in the little town, a horse, and a cow.

Most people in the Pale of Settlement knew oppressive poverty, and epi-demics of tuberculosis and other diseases were common. The Rosenblooms lived from hand to mouth. The younger children wore hand-me-downs. There were six daughters and no sons, although Vulf very much wanted a son. Anna, the second born, was a very capable young girl and was noticed by a local teacher, who, against the odds, succeeded in getting her a free education in a ministerial school; she finished the five-year program with a certificate of distinction. From the age of 12, Anna worked as a dress-maker to help the family. She remained intellectually curious. On one of his trips, Zhores recalled hearing, Vulf found, in a forest, a box of books by Pushkin, Lermontov, Tolstoy, Turgenev, and Dostoevsky, and brought them home. Anna read them all by candlelight, told the other girls about them, and, years later, entertained Zhores and his brother Marx by recalling entire passages from memory. Zhores believes that his lifelong love of literature comes from his mother.

Jews had a complex and contradictory life in the Russian empire. More urban than many other minorities yet largely restricted to villages, they were stuck in lower-class artisanal and trade positions and wage labor by law, especially after the freeing of the serfs in 1861, yet they found many representatives among wealthy financiers and industrialists, especially in mining, railroads, and transport. Though limited in educational opportuni-ties by quota, they made up an increasing percentage of university students. Their legal disabilities, and the pogroms, led many to emigrate (2 mil-lion of them to the United States) and others to move to urban centers.[18]

Younger, more urbane Jews turned to Marxism as a means of revolt against Jewish tradition, against their parents, and, of course, against capitalism. Many of them rejected Yiddish for Russian as their first lan-guage. They joined populist and terrorist political parties opposing the tsarist regime, seeing socialism as emancipation from rural backwardness and capitalism. They became prominent in the Bolshevik Party, both in leadership positions in the Central Committee after the revolution and in the Red Army, two-fifths of whose high officials were Jews. Interestingly, much of the civil war was fought in the old Pale of Settlement, with Poles,

Ukrainians, and Whites seeing the Bolsheviks—and the Jews generally—as both aliens and enemies.[19]

Though Zhores Alferov claims never to have experienced anti-Semitism, it was ingrained in Russian society, and it grew worse during the Stalin era. It led to Stalin's infamous campaign against Jewish doctors in the late 1940s and the early 1950s (see chapter 2), and it found fresh energy in the Brezhnev era, when Zhores's Jewish colleagues faced discrimination, even though Soviet Jews tended to be, like Zhores and scientists generally, urban and well-educated. Because of the experience of Zhores's mother and other relatives under the tsars, during Soviet power, and of course during the Nazi invasion, it is important to discuss briefly how Jews lived in Russia and in Belarus.

Before the Russian Revolution, the reactionary anti-Semitic "Black Hundreds" and ordinary people carried out pogroms in Moldavia (today called Moldova), in Ukraine, and in Belarus. Zhores Alferov recalls learning that Belarus was immune to these attacks. Even in poverty-stricken Kraisk, he believes, Belarusians did not engage in anti-Semitic violence. In actuality, Anna Vladimirovna's family was at the center of violence directed against Jews under the tsars and of the effort to exterminate them during the Nazi Occupation. The tsarist government, both anti-Semitic and fearful of revolution, used the Okhrana (secret police) to make matters worse for Jews. The Okhrana fabricated the infamous and incendiary Protocol of the Elders of Zion, a document allegedly providing proof of the worst lies about the Jews—blood murders, international conspiracies, and so on. Using informers and agents, the Okhrana stirred up anti-Jewish sentiments and fed the fires of pogroms, especially in regions of what is now Belarus (Vitebsk, Gomel, Mogilev). Whenever things went poorly for Russia, citizens found new reasons to blame non-Russians. For example, the failed Russo-Japanese War provoked pogroms in 1905, as if the Jews or Jewish soldiers were somehow to blame. The tsarist authorities built on the discontent of the local population, even printing scurrilous leaflets that attacked the evil "Yids." Pogroms occurred after the Russian Revolution too.[20]

The Nazi invasion of 1941 resulted in the death of nearly all remaining relatives of the Rosenblooms. Others had already been killed. Zhores's mother had two sisters, Manya and Rosa, who lived in Pleshchenitsa in the heavily Jewish region of Logoisk, 60 kilometers north of Minsk. Before the war they both had sons, both named Volodya, both born in 1932. Manya's husband, the chairman of the local agricultural Soviet, was arrested in 1937 and shot for imagined crimes. The German armies reached Minsk by June 28, 1941, and Zhores's aunts were unable to evacuate in time. While many

Ukrainian and Belarusian peasants collaborated with the Nazis, others, at great risk to themselves, hid Jewish families in their cellars for weeks at a time, moving them from house to house. When the partisans (pro-Soviet guerilla bands) first appeared in the region, they not only attacked German soldiers but also tried to protect Jews; approximately 10,000–20,000 Jews joined them. Aunt Manya married a partisan soldier, Mikhail Vladimirovich Savelzon. Savelzon had worked as the director of a forestry cooperative before the war, then, like Manya's first husband, had been arrested in 1937 as a "wrecker" and an "enemy of the people" and sent to remote, isolated, brutally cold Kolyma on the northern Pacific coast. For most prisoners, Kolyma was the equivalent of a death sentence; they were forced to toil in mines and the forests with rags as clothing and shoes, to live on miserly rations, and to dig for gold and uranium in –50°F temperatures with only such rudimentary equipment as picks and shovels. During his interrogation, Savelzon never admitted guilt. Though not confessing did not guarantee survival, in this case it did not hurt. In 1938, Stalin's Party, in a fit of temporary concern over its murderous policies, allowed a brief wave of rehabilitation of those wrongly accused, and Savelzon was released and sent home to Belarus. He lost his wife and children to the Nazi advance in the first days of the war, but he found new happiness with Manya. In the early fall of 1945, Aunt Rosa returned to Pleshchenitsa, but on the way home she fell out the door of a truck, hit her head on a bridge approach, and died on the spot. Having survived the entire war, first literally living underground and then joining the partisans, she died in a time of peace. Her son Volodya was then adopted by Aunt Manya and Uncle Misha (Mikhail), who, as was mentioned, had a son of their own by the same name.

When the Nazis occupied the USSR, perhaps 4 million Jews lived within its borders. By the end of the war, 3 million Jews had been murdered. Perhaps as many as 50,000 Jews joined the Red Army during World War II but hid their identities to avoid Nazi terror if captured. Of the 3 million Jews who were murdered, 800,000 were Belarusian. Belarusian collaborationists played a role in murdering them. Though the Germans undoubtedly were the main driving force behind the attempt to eliminate all Jews, the number of the victims indicates that local residents also participated in the exterminations. This may be because, although Jews and gentiles had long lived together in the Pale of Settlement and had been forced by economic circumstances to work together, many Belarusians, Ukrainians, and Russians had always seen Jews as an alien element.[21]

Collaboration reflected popular sentiments and helped the Nazis accelerate the killing. Pogroms became common again, and Jews not killed by

Nazis or their supporters often were forced into ghettos, where they were killed later. Minsk had one of the largest ghettos; it was destroyed in 1942 after a revolt broke out. Jews were exterminated in the ghettos of Vitebsk, Polotsk, and Glubokoje too.[22] After Vitebsk had been rid of German troops, only 118 out of 180,000 residents remained, and the city lay in ruins.[23] As for relatives on Alferov's mother's side, of six daughters, three died in the war, and, as has already been noted, Rosa died at the end of the war after falling out of a truck. Zhores's cousin Vladimir emigrated to Israel. No relatives remain in Belarus, although the cemetery in Chashnika has grave-stones of aunts and grandparents.

During World War I, Kraisk had fallen under German occupation, then Polish occupation, then into the maelstrom of the civil war between Reds, Whites, Cossacks, and various other political and ethnic groups and anar-chists. The border between Poland and Belarus followed a creek that flowed along side of Anna's house on the outskirts of Kraisk. This meant that Kraisk was at the center of military action, smuggling, and violence against Jews. Anna's facility with foreign languages saved her and her home on one occasion. She knew German and Polish. In 1920, White Polish soldiers entered Kraisk during a pogrom. One officer barged into the house, intend-ing violence. Anna spoke with him calmly in Polish. Assuming that she was not a Jew, but perhaps a Polish employee, he left the home untouched.

Ivan ran into similarly unexpected dangers in Kraisk. During the civil war, both borders and people's allegiances shifted, depending upon whether Reds or Whites controlled a territory. Bands of smugglers and criminals moved freely in conditions of anarchy. Armed soldiers of the no-torious Bulak-Bulakhovich Gang crossed the frontier frequently, stealing everything not nailed down, killing communists and other activists, and carrying out pogroms in the towns and villages.[24] This group and Whites in the region were eliminated only in November of 1920, under Trotsky's leadership. For Ivan or any other Chekists to confront this dangerous situa-tion presented great risks, but one that Ivan apparently handled well, win-ning recognition and promotion by Party officials.

The young Chekist Ivan Karpovich Alferov encountered a warm envi-ronment in the Rosenbloom home. In Anna, Ivan found not only an at-tractive, educated woman but also, Zhores writes, "a friend and comrade who shared his ideals and political conventions." The two young people fell in love and determined to marry, despite the misgivings of Anna's fanatically religious mother, Khaia. (Ivan was Russian Orthodox.) Vulf sincerely loved his son-in-law. They had become close friends, perhaps because both had been soldiers and knew the human costs of war. Vulf supported the marriage, but Anna's mother never forgave her for marrying a gentile.

In 1922, Ivan agreed to be transferred to Starobin, a shtetl on the Sluch River about 150 kilometers south of Kraisk. On a dark, rainy November night, while Anna's mother, aunt, and younger sisters slept, Vulf helped Ivan and Anna load their belongings onto a cart, and they set off for Starobin. Ivan and Anna rented a room in the house of a Roman Catholic priest. The priest apparently worked for the Polish intelligence service, but Ivan was able to gain his confidence while playing cards with him. The Alferovs remained in Starobin only a short time; in 1923 Ivan was demobilized and was transferred to Polotsk, a city on the western Dvina River in Vitebsk province at its confluence with the Polota River, for service in the nascent customs bureaucracy.

The Rosenblooms lived in a region of great literary activity. Zhores's mother and Zhores himself came to revere Belarusian literature. Zhores's facility in Russian, Belarusian, and eventually English distinguished him from many Soviet citizens and many of his future colleagues. Scientists might learn to read German or English, but few had the opportunity to use it abroad, and fewer still sought the opportunity to embrace the study of world literature in the original, let alone Belarusian literature. Although poorly known even in the former Soviet Union, Belarusian literature had a great tradition that flowered briefly in the 1920s. One of the leading writers, Zmitrok Biadulia (Samuel Efimovich Plavnik, 1886–1941), lived in Kraisk. Zhores's mother knew the poet Ianka Kupala (Ivan Dominikovich Lutsevich), whom she met through Biadulia. At least until the rise of Stalin, Bolshevik influence over Belarusian cultural institutions was limited. Even after the devastation of war and revolution, Belarusian specialists in all areas created a series of important cultural institutions—the Scientific Research Institute of Belarusian Culture (later the Belarusian Academy of Science), the Belarusian University, and pedagogical, medical, and technical schools. Several literary associations were founded, many of which published literary magazines. One association, Maladniak (Youth), had a membership of 500 writers.

Zhores Alferov received his love of culture not only directly from his parents, especially his mother, but also from his understanding of the environment in which his parents lived in the 1920s, at the epicenter of flourishing Jewish and Belarusian secular life. One of the centers of Jewish life before and after the revolution was Vitebsk, the bustling industrial and cultural city in the Pale of Settlement in which Anna and Ivan lived for a short time. In spite of the occupation of part of Belarus by Polish troops in the first half of 1919, Vitebsk acquired a symphony orchestra and a conservatory. The city had three theaters, six clubs, and several movie theaters.[25] The critic and philosopher Mikhail Bakhtin and the painter Marc Chagall lived there. After the revolution, the architect and artist El Lissitzky (Eleazar Markovich Lisitskii) worked at the Vitebsk art school, where he met Chagall and Kasimir Malevich.[26] Malevich, with Piet Mondrian a founder of modern art using geometric forms, pursued his new art form based on abstract geometric patterns—Suprematism—in Vitebsk.

Children of the Russian Revolution

On January 1, 1924, Anna Vladimirovna Rosenbloom gave birth to a son, whom the parents named Marx Ivanovich Alferov. The couple and the child visited Anna's father in Kraisk, but Anna's mother refused to see the new parents or the baby boy.[27] Bolshevik propagandists, on the other hand, paid close attention to this birth and those of other children. The birth of a Soviet child became the object of a ceremony that they used to wean the people from religious belief. On January 13, the Alferovs' son received his "red baptism," which meant he became an honorary member of a local trade union. The "red baptism" usually took place in a Party headquarters. The secretary who conducted the service supplanted the priest, while the workers' council of the factory in which the father was employed served as the godparents. Often a choir of workers sang the *Internationale*. When a curtain rose, parents and children sat together around a table on the stage. A speaker explained the meaning of the ceremony. The fathers then stood, each expressing how he would raise the child in the spirit of the new revolutionary Russia.[28] Marx received a copy of Marx's *Capital* inscribed by the workers of the Polotsk customs office Party cell "To young comrade, Marx Ivanovich Alferov"; it was presented in the spirit of freedom from "all religious prejudices" and on the understanding that "education along the right path . . . leads to socialism."

In 1926 the Party apparatus sent Ivan Karpovich Alferov to Gorodok, a village not far from Vitebsk, where he worked as deputy chairman of the regional executive committee, then as director of a sawmill. The general recovery of the Soviet economy assisted Ivan in getting the Gorodok sawmill to operate profitably, and he stayed there for two years.[29] The New Economic Policy (NEP) created the conditions for recovery. In 1921, at the tenth Party Congress, Lenin pushed for abandonment of War Communism and urged Party members to approve the NEP. War Communism had failed to resurrect the economy; indeed, production plummeted in virtually all

Zhores Alferov's parents, Ivan Karpovich and Anna Vladimirovna, in Gorodok, Vitebsk Province, 1927, where Ivan served as a border guard for the Cheka, the secret police.

sectors of the economy, a dreadful inflation swallowed any savings, and harsh militarization of labor and confiscation of grain at the point of a bayonet alienated workers and peasants respectively.[30] During the NEP, which Lenin imagined might last as long as 25 years, the government maintained control of major and strategic industries while permitting small-scale trade and a money economy to flourish. Lenin anticipated that the provision of tractors would result in a gradual conversion of the peasantry to socialism. By 1926, the economy reached 1913 levels of production. The entrepreneurs of the period, called NEPmen and brilliantly portrayed as hucksters and swindlers in Ilya Ilf and Evgenii Petrov's novel *The Twelve Chairs*, contributed mightily to economic recovery as suppliers of scarce goods.

Zhores Ivanovich Alferov was born in Gorodok on March 15, 1930. His parents had anticipated a girl and had thought of naming her Valeria. Ivan selected the name Zhores in honor of the French socialist Jean-Joseph-Marie-Auguste Jaures (1859–1914), one of the leaders of the French Socialist movement during its formative years. Jaures attended the Ecole Normale Superieure in Paris, earning a philosophy degree, and later (1883– 1885) lectured at University of Toulouse. He served in the parliament intermittently. He became a socialist in the 1890s, on one occasion losing a reelection campaign because of his support of the innocence of Captain Alfred Dreyfus. Jaures espoused a view of gradual adoption of democratic socialism rather than violent revolution, although as an advocate of the Second International he refused to participate in "bourgeois" governments. He co-founded the socialist newspaper *L'Humanite* in 1904, and under his leadership the French socialists united in 1905. He was assassinated on the eve of World War I.[31]

In Gorodok the young family acquired a foxhound and named him Kolchak (after Alexander Kolchak, a tsarist officer who served as Minister of War and the Navy for an anti-Bolshevik government in western Siberia through 1919).[32] When the family received the first of many distant postings to Permilovo on the Vaimuga River in Arkhangelsk province, Anna decided to part with Kolchak. She sold him several times, and each time he escaped from his new masters to return to the Alferovs. Anna jokingly wrote Ivan, who had already settled in Permilovo, "You don't need to send money. I'll continue to sell Kolchak and live on those funds." Kolchak disappeared from his last owners in search of the Alferovs, never to be seen again.

In Permilovo, Ivan continued to work in pulp and paper mills. The dense forests of Arkhangelsk province (north of Moscow and Petersburg) served as one of the major sites for the pulp and paper industries until after

World War II, when, because of clear cutting near transportation routes, the industry began to move to the taiga in the Urals and beyond to Siberia. Because of the swampy, boggy land, the dense forests, and the very low population densities, tsarist entrepreneurs and engineers found it difficult to build roads and railroads in Arkhangelsk province. Peasants lived in poverty, ever on the edge of crisis. But the lumber industry was able to use the huge network of lakes and rivers as a major mode of transport during the spring, when rivers thawed.

It fell to Ivan Karpovich Alferov's comrades in Party organizations to organize the peasants into collective farms and lumbering operations, which would be supplemented in the Stalin era by the forced labor of tens of thousands of prisoners in the gulag (the Russian acronym for Main Administration of Corrective Labor Camps). The peasants—and the prisoners—hated the work, were paid and fed poorly, and lived in hastily constructed barracks. They had to rely on brute human force rather than rudimentary machines. Owing to poor harvesting practices, inadequate mechanization, and under-motivated workers, the industry remained inefficient and even criminally wasteful in its use of lumber. Perhaps 40 or 50 percent of the timber rotted on the forest floor or sank in rivers. Eventually the prisoners built forest roads and a few narrow-gauge railways to the pulp and paper mills from the forestry collectives. Today Arkhangelsk's mills are its major producers—and polluters. Dioxin and other harmful wastes have been dumped haphazardly into the environment for 80 years, and Arkhangelsk's water supply is one of the most unsafe in all of Russia.

A Proletarian Family in the Early Stalin Era[33]

Ivan Karpovich Alferov's responsibilities to the state and the Party grew significantly with Stalin's rise to power. Stalin and the militant communists who came to power with him had grown impatient with the slow pace of "socialist reconstruction." They rejected the NEP as an abandonment of socialist mores, not to mention socialist economic institutions based on state ownership of the means of production. They called for rapid industrialization and collectivization of agriculture—the latter of which Lenin had rejected as premature because of the peasants' lack of receptivity to it. Stalin also adopted a radical view of cultural change. Rather than rely extensively on "bourgeois specialists" (meaning scientists, engineers, and managers whose careers dated back to the tsarist era), the state would train reliable workers and deploy them throughout industry, agriculture, and research. In many ways, the Russian Revolution reached its conclusion with the programs that Stalin advanced as a Great Break (velikii perelom) with the remnants of capitalism in Russia. And, of course, Stalin's rise to dictatorial power accompanied the Great Break.[34]

Stalin advanced five-year plans to "reach and surpass" the West in industrial production in a matter of years. Zhores Alferov remembers hearing of lathes that were called "DiP100" and "DiP 200," DiP being the acronym for the Russian words meaning "reach and surpass." He also remembers hearing that Stalin's crony Viacheslav Molotov, a diplomat who also was chairman of the Council of Ministers, had warned "Maybe we oughtn't surpass; they'll see our naked behinds."[35] In any event, the fervor to surpass created an atmosphere of heroic accomplishment even as laborers toiled to boost output in every industry with outdated technology under miserable conditions. Individuals who spoke of a more gradual approach and of paying more attention to the plight of the worker during the industrialization campaign, or whose enterprises failed to reach plan targets, were often treated as "wreckers." The first show trials in the USSR were a kind of public theater to used to skewer scientists and engineers, many of them foreign, in order to silence moderate voices.[36]

As part of a propaganda campaign to exhort workers to aim at higher targets, the authorities engaged in a public relations campaign. Ivan Karpovich Alferov's work at the Permilovo mill was so successful that the authorities produced a film about the victory of its workers over nature (a film I have been unable to find in the archives). The director Viktor Turin produced a documentary film, *Turksib* (1931), on an early "hero project," the Turksib (Turkestan-Siberia) Railway, intended to link the grain-surplus areas of Siberia with the grain-deficit but cotton-rich territory of Turkmenistan.[37] Ivan's successes as captured in film led, in 1932, to another posting not far from Arkhangelsk—a posting that enabled him simultaneously to study at the Arkhangelsk Industrial Academy, one of many such institutions founded in support of expanded lumbering efforts.

Party leaders signaled their intention radically to transform the relationship between specialists and the state in two ways. The first was a sinister and violent campaign to subjugate the scientific and engineering community to state economic-development programs. It was directed physically at specialists whose training dated to the tsarist era (whom Communists had come to believe were inherently untrustworthy as representatives of the bourgeois social order), and psychologically at all future specialists. In 1928 and 1930, Soviet prosecutors held two public show trials in which they accused engineers of being "wreckers" and/or spies working on behalf of foreign powers. The trials culminated the investigation of the so-called

Shakhty (Mining) and Industrial Party Affairs. The prosecutors jailed a large number of specialists, and later executed a number of them, sentencing others to lengthy prison terms, although the German mining engineers accused in the Shakhty Affair were eventually released. For most scientists and engineers the lessons were clear: Do not engage in politics, do not fight young Party bosses over the appropriate role of scientists in Soviet society, and do not argue with the plan which should take precedence over any scientific consideration over what was rational or possible.

The second radical transformation involved a new approach to education, the advancement of workers, and the expansion of the higher educational system that accompanied the first five-year plans. Ivan Karpovich Alferov gained advanced training in the early 1930s as part of a nationwide movement to create a class of communist specialists. The desire to advance young communists quickly accompanied the industrialization effort and proceeded with the same tempo, urgency, successes, and bottlenecks. Party officials intended no less than a cultural revolution to transform the class makeup of students through a program of "advancement" (vydvizhenie).[38]

Through their experiences of the civil war, and through service to the Party, Ivan Alferov and Communists like him had already demonstrated their suitability for the process of advancement. To train these individuals, officials in industrial and educational ministries founded a series of institutes, one of which was the Arkhangelsk Industrial Academy.

Yet vydvizhenie failed to train qualified engineers. Too many of them lacked even basic reading, writing, and mathematical skills. They often worked at the same time that they studied and thus were exhausted by the nearly 24-hour effort to advance themselves and industry simultaneously. Their "institutes" were often little more than rooms and blackboards. By 1931 vydvizhenie had to be scaled back, since the graduates from workers' faculties ("rabfaks"), technical schools, and other institutions simply could not handle all the science, the technical subjects, the Party history, and the Marxism that they had to master in the short time that was allotted them.[39] Many of them had returned to the factory with a certificate in hand, but without the requisite skills, and in some cases they had unintentionally damaged equipment or slowed production. Still, by the end of the 1930s the Party had introduced an entirely new system of higher educational institutions, ranging from vocational institutes to universities. These institutions turned out tens of thousands of members of the new Soviet intelligentsia, including many if not most of the directors of factories and virtually all of the Party leaders who would assume power in the 1950s and later. They also turned out such industrial managers as Ivan Karpovich

Alferov, who became a leading manager in the forestry, pulp, and paper industries.

In 1932, Ivan entered the Industrial Academy (sometimes called Promakademiia). He took courses in physics, higher mathematics, inorganic chemistry, drafting, quantitative and qualitative analysis, economic geography, the German language, the Soviet economy, and political economy. He was among the elite of young communist students of the proper social background, and he got good grades in every course.[40] With his three years of elementary schooling, he had no difficulty meeting the demands of courses taught at the middle-school level. No doubt his long experience in the industry and his knowledge of its technology made him a good choice for the Academy. His 1935 diploma listed his qualifications as an "engineer-organizer" in the pulp and paper industry, not general training in management such as an MBA candidate might get today. Upon his graduation from the Industrial Academy, he was appointed director of a recently built pulp and paper combine at Solombala, just across the Kuznechikha River from Arkhangelsk proper.

At such higher educational institutions as the Promakademiia, in addition to technical subjects, students received instruction in Marxism-Leninism, Party history, and so on. They were joined to such national campaigns as socialist competitions to increase production and Stakhanovism (named after a Ukrainian miner who overfulfilled his norms by superhuman quantities).[41] They took on "socialist obligations" in areas ranging from study to production to ideological purity. The Promakademiia was noteworthy for the fact that, as far as can be judged from archival documents, all 87 of the students in its first years were Communist Party members. Fifty-five of them were from working-class backgrounds, 16 from peasant backgrounds, and 16 from white-collar backgrounds. Of the 29 professors and instructors, seven were Party members and all had earned degrees or titles in the Soviet era.

In the spring of 1934, the Industrial Academy's faculty and students participated in an All-Union Socialist Competition of Higher Educational, Higher Technical Educational, and Vocational Schools. At the Promakademiia, as part of the competition, students and faculty members promised to carry out and publicize scientific research through "theoretical evenings" and publication in journals of the forestry industry. Their topics included "the ideological bases of wrecking (vreditel'stvo) in forestry" and Karl Marx's reputed contribution to forestry (which must have gone beyond the trees sacrificed to publish his 25 volumes of works in tens of thousands of copies, and countless pamphlets). Fully 95 percent of the Promakademiia students

The Alferov family in Arkhangelsk (where Ivan worked in the pulp and paper industry), Zhores standing in back and Marx sitting in front, 1934.

gained the honor of being named udarniki (shock workers). Throughout the late spring, on the campus and throughout the region, the competition called forth a series of no doubt stimulating lectures and seminars, as well as field trips. The students visited a lumber mill, a tractor-repair facility, and the Arkhangelsk Electrical Power Station (whose production always lagged behind the growing demand, in part because the authorities directed lumber that might have been used to fire the boilers to the pulp and paper mills). The students saw two films in connection with a course on internal-combustion engines, but most of their "competition" concerned classroom activities on assembly and repair of forestry equipment, establishment of technical norms, improvement of delivery of goods and services, hydrolysis, electrotechnology, and forestry itself.[42]

Zhores's mother, Anna, quickly tired of Arkhangelsk, which she found dreary, industrial, and cold. The housing and food situations were particularly trying. Housing was very tight throughout the nation, and during the collectivization campaign agricultural production dropped and deliveries to the cities fell sharply. Bread was in short supply; lines for it and for some other goods began to form in the middle of the night in spite of the vigilance of the secret police. As a student at the academy, Ivan was given one room in the dormitory. He chose a room by the tiny kitchen, but Anna always remembered it as a poor choice because of the noise and smells and the boiler below that banged through the night. The ration-card system enabled the family to live only a "half-starved" existence, Zhores recalls, while millions of peasants in Ukraine starved to death during the violent collectivization campaign. They lived on boiled potatoes and tea. In time for various holidays they often received food parcels from the grandparents on both sides of the family, whose gardens produced a modest surplus. They sometimes shared these parcels with another family in the dorm. On one occasion, the three-and-a-half year old Zhores saw some boiled eggs on the table and, not knowing about the shell, tried to devour one whole.

As a distraction from the hard life of Arkhangelsk, Zhores's mother entered a "rabfak" to improve her qualifications. A student might earn a kind of high school equivalence degree at a rabfak, then enter a regular institution of higher education. Hundreds of rabfaks throughout the nation helped thousands of proletarian students improve their qualifications. Anna became a librarian and an avid volunteer in all sorts of social organizations.

Marx Alferov entered elementary school in Permilovo in 1931. Because he was home after mid-day, his brother Zhores spent a lot of time playing with him and his friends. On one occasion the boys amused themselves in the yard of the academy, repeatedly jumping over a water-filled ditch. The older boys cleared the ditch, but Zhores fell into the deepest part; Marx quickly pulled him out before he drowned. Marx also saved Zhores once when he was playing on tram rails, pulling him out of the way of a tram just in time. Zhores wanted not only to play, but always to be the leader in play, so he got the nickname Narkom (meaning People's Commissar). When he got bruised or cut, a neighbor might cry out "Anna Vladimirovna! Come get your Narkom. He's fallen down."

Like many small boys, Zhores was a bundle of energy and played hard. He also displayed several qualities that would serve him well in later life. He understood complex issues fairly comprehensively for a boy his age, and he had an astounding memory. Hearing a poem about the work of a secret underwater expedition and the sinking of the ferry *Sadko*, he memorized it on the spot and then pretended to read it from a book. This provoked shock and appreciation among adults. In February of 1934, the Communist Party held its seventeenth congress with a joyous—and exaggerated—celebration of achievements in meeting the targets of the five-year plans ahead of target dates (a lie), of a triumph in collectivization in agriculture (true, but ignoring the famine that resulted in deaths of millions or the slaughter by the peasants of half of Soviet livestock rather than enter the collective farms), in machine building, and in metallurgy, and with lengthy discourses on high unemployment in the West (true) and on the danger of capitalist encirclement. By the time of the Great Terror (1937–38), 65 percent of the delegates to the congress had been purged—arrested and executed—on Stalin's orders. But for young Zhores Alferov and for most other citizens, the congress indicated that the country had moved closer to socialism under Stalin's infallible leadership. Zhores memorized sections of the speeches of Stalin and those of Kliment Voroshilov, an old Bolshevik and a military leader of the 1930s and the 1940s who not only survived the purges but took advantage of them to advance his own position.

During a brief vacation in the summer of 1937, Ivan worked on the construction of Molotovsk, a shipbuilding city that was being built largely by prisoners. (Renamed Severodvinsk in 1957, the city drew Zhores Alferov in the late 1950s to consult on the development of solid state circuitry for the first generation of Soviet nuclear submarines.) Ivan Karpovich Alferov could not have avoided contact with political prisoners and forced labor, either in the lumber industry or in Molotovsk. Zhores gives the impression that his father was troubled by this aspect of "socialist reconstruction" but was not a party to it, and that to avoid recriminations or arrest he kept his mouth shut.

In 1937, Ivan Karpovich Alferov was transferred to Stalingrad to assume an appointment as director of a woodworking combine in under the jurisdiction of the Commissariat of the Means of Communication. Many of the factories of the woodworking industry fell under the Commissariat of the Means of Communication at this time, since the rivers and railroads were the main arteries for the transport of lumber. (Later they were transferred to the newly established Commissariat of the Pulp and Paper Industry.) Anna was quite happy to be out of Arkhangelsk. On the way to the new posting, the family spent about ten days in Moscow at the apartment of Ivan's godfather, Abram Solntse. Alferov kept a copy of Kipling's *Jungle Book* that Solntse gave him as a gift, while Solntse gave Marx a book that glorified the political reeducation of homeless youths during the NEP period. The family arrived in Stalingrad in summer. The factory was located in the small town of Beketovka.

Perhaps Ivan's promotions were attributable to his work ethic. Every evening he inspected the Stalingrad factory, and Marx and Zhores often accompanied him.[43] For the first two months or so they lived in a communal apartment with two small rooms in a workers' village called Kozii Khutor;

then they were given a house that had belonged to the woodworking mill's owner before the revolution. It had been turned into a duplex. The secretary of the Party committee of the factory occupied one part, the Alferovs the other part. This one-story cottage stood on the shore high above the Volga River and was surrounded by a garden. The yard had a small fountain, an outdoor summer kitchen, and a pasture for animals. Zhores thought it heaven in comparison with the dormitory in Arkhangelsk. Though the family still lived on ration cards, tomatoes and melons—which Zhores had never seen—were already coming in to the local markets; the region had rich vegetable and fruit farms.

Anna was active in social-political organizations. She joined a kind of proletarian "junior league"—the Soviet Zhen-Obshchestvennits (Wife–Social Activist) movement—that was dedicated to fixing up the workers' dining hall, to kindergartens, and to home life. In Beketovka the organization succeeding in establishing a municipal park—the authorities were determined to make Soviet cities "greener" than their capitalist counterparts. In 1936, Anna was elected a delegate to the All-Union Congress of Wife–Social Activists, and the family journeyed to Moscow for the meeting. The travel agents put Anna in a magnificent room in the Grand Hotel, while Ivan stayed in the new Moscow Hotel. The boys spent ten days with their father's friends in Losinoostrovskaia, a beautiful suburb of dachas owned by privileged officials. At the congress, Anna's presentation so impressed Lazar Kaganovich (one of Stalin's leading henchmen, sometimes head of the Moscow Party organization, and a major force behind the construction of Moscow's subway system) that he awarded her a beautiful tea service, some of the pieces of which Zhores Ivanovich and his wife, Tamara Georgievna, still have at their home in Komarovo.[44]

The return trip took the Alferovs by train from Moscow to Gorky (Nizhnyi Novgorod), and by steamboat from Gorky to Stalingrad. The Moscow-Volga canal—one of Stalin's great "hero projects" of the 1930s, and part of the "socialist reconstruction" of Moscow—was not yet finished, but evidence of public works construction projects was everywhere. As Zhores tells it, one evening the family dined in a restaurant at the Gorky River Port. At the next table sat a lieutenant of the Red Army. At that time, Red Army commanders commanded universal awe and respect. Zhores had made an impression with his poetry recitals, so the commander invited him to sit with him and treated him to ice cream while indulging himself in another glass of wine. Zhores was deeply moved that a tall and handsome officer had taken interest in him. The lieutenant excused himself for a moment to visit the men's room. But he never returned, and Zhores's father had to pay for the dinner and the ice cream. For a long time Zhores had nothing but bad memories of Gorky. The remainder of the trip on the steamboat, with stops to stroll in Kuibyshev and Saratov, was blissful. The boys spent most of their time on deck. Marx already loved the water and spent the summer in the Volga swimming back and forth from their region of Beketovka to a huge island in the center of the river across from Akhtuba. The family ate well, as fishermen sold fresh starlet and sazan, from which the Alferovs often made soup.

Zhores and his family always remembered Stalingrad fondly, both for the quality of life and for the fact that Ivan's career continued its upward trajectory. On his arrival, Ivan reported to the Stalingrad territorial Party Committee to present his papers. The first Party secretary was an old Party acquaintance, I. M. Vareikis. Ivan was given an automobile and a driver, although as a former Cheka and cavalry officer he preferred to use his horse.

Purges and Party Faithful

For the Alferov family, the Stalingrad years ended as the Great Terror unfolded. During the Great Terror, Stalin and the NKVD chiefs Nikolai Yezhov and Lavrenty Beria unleashed an assault on members of the Party elite, on leading managers and planners, and on Party intellectuals. Yezhov had demonstrated his devotion to the Bolsheviks in a long career before assuming control of the Great Terror. He joined the Bolsheviks in Vitebsk in May of 1917, served in the Red Army, and later moved into the Commissariat of Agriculture as deputy commissar during the brutal collectivization campaign that amounted to war against the countryside. Late in 1938, Beria replaced him, and Yezhov was executed in 1940. Under Yezhov the secret police arrested such Old Bolsheviks as Nikolai Bukharin, Mikhail Tomsky, and other associates of Lenin and accused them of unspeakable and untrue crimes of terror and espionage against the state. The military also was purged. Hundreds of thousands of people were arrested, and tens of thousands were shot.[45] No one was safe.

One day in 1938, Ivan Karpovich Alferov went to Moscow on business.[46] A few weeks later, his wife received a letter saying that he had moved to Novosibirsk and was at work at another pulp and paper facility. He asked the family to pack their things and follow him to Siberia, and so they did. Only after Khrushchev's "secret speech" at the twentieth Party congress in Moscow (1956), in which he denounced Stalin's murderous "cult of personality," would Ivan explain how he ended up in Novosibirsk. When he arrived in Moscow, the head of the Main Administration of the Paper Industry, an old comrade, advised him not to return to Stalingrad. The local secret police had uncovered an imaginary conspiracy involving the entire leadership of the Stalingrad factory. The directors had been identified as "enemies of the people," an appellation that usually meant impending execution. The Moscow official instead signed orders making Ivan deputy director of the Western Siberia Forestry Trust (ZapSibTransLes), where, indeed, another conspiracy led to the dismissal of all the ZapSibTransLes directors anyway. He convinced Ivan to go to Novosibirsk, assume his new post quickly and quietly, and send for his family later.

It helped that, in spite of their vigilance and their reputation for efficiency, the various regional and local secret police offices had poor connections. They might fulfill their plan for arrests in Stalingrad, but could not know to search for Ivan Karpovich Alferov in Siberia. Hence, Alferov and his family were saved temporarily from the charge of being "enemies of the people." But in the weeks before Ivan sent for his family, Anna and the two boys surely worried about the loss of husband and father. During the Great Terror, when someone disappeared his relatives expected the worst. And most people did not have any connections that might save them from arrest.

Anna and the boys arrived in Siberia in August. As usual, there was no apartment ready, so they stayed in a hotel. As was his habit, Ivan arose early and returned late, when the boys were already asleep. Anna filled time with them with a variety of card games. Not knowing where their apartment would be, Anna registered the boys for middle school in a nearby region. Marx entered seventh grade. Anna wanted Zhores to go to school, although he was a year younger than the other first graders. At first he refused. Anna insisted he must, since otherwise he would be in the hotel by himself all day. He fought her and cried, but eventually agreed to start school.

Zhores had the good fortune of having Maria Mikhailovna Sosunova as his teacher. She had finished the Bestuzhevskie courses in St. Petersburg (a special university program established for women, the first of its kind in Russia, dating to the 1870s), and as a young girl had moved to Siberia to teach. She was at least 40 years old, and of large dimensions. To Zhores she was an "unusually lively old diva." She taught first grade her entire life. She made things interesting. First grade did not run the entire day. But, having gone home and had lunch, the first graders often returned to the school to be with Sosunova, do homework, and play. Sosunova lived in an apartment belonging to the school and was entirely devoted to the children.

Zhores had the first of several misfortunes in play that year. He had developed a series of experiments with explosives. In one experiment, the mixture blew up in his face and he lost his sight for two months. Sosunova and his classmates visited often to keep him up in spirits and in his studies. When the Western Siberia Forestry Trust finally gave the Alferovs an apartment, but in a distant neighborhood, Anna intended to move Zhores to another school closer to home. He did not wish to move. Anna asked the teacher for her support in the move, but Sosunova supported Zhores as being already "a real person," so he remained at his first school. When the family moved to the Urals region during the war, Zhores read in a newspaper in 1944 that Sosunova had received the Order of Lenin, the nation's most prestigious award, and sent her a letter of congratulations.

One day, Ivan Karpovich Alferov did not return from work. Anna informed the boys that he was on a business trip up north. At about 3 a.m., fearing the worst, she went to the trust offices to learn what had happened. The trust was in a huge building that also housed a number of other offices. The night watchman inquired who she was, then sadly informed her that all the administrators had been arrested. Alferov was accused of far-fetched crimes that might easily lead to exile or execution. The NKVD (People's Commissariat for Internal Affairs) contended that he had wasted resources in the feeding of workers. The Western Siberia Forestry Trust had a large number of cooperatives spread through the region, and a large number of prisoners worked in them. The prisoners were supposed to be fed at the same level as other workers. The NKVD accused Alferov of feeding political prisoners more than common criminals. Like hundreds of thousands of others, Alferov admitted no crime and signed no confession. In 1938, Stalin, in a strange retreat into justice in light of the maelstrom of the purges that he himself had triggered, signed a proclamation that required a procurator to countersign interrogation documents. Fortunately, the procurator in the western Siberian region had also been in the cavalry during the civil war. He believed that Alferov was a decent communist and knew that he was a civil war hero. He must also have understood that there was no basis for the arrest. Alferov was released some weeks later and returned to work as if nothing had happened, but he was very thin and for a while he had a very sad look on his face. His release probably was attributable to the arbitrariness and irrationality of the purges as much as to the prosecutor's temporary decency.

Even after his arrest and release, Alferov continued to respect Stalin and the communist ideals he believed Stalin represented. Like hundreds of thousands of others, he believed that his arrest had been a mistake, that the authorities would straighten things out, and that the others might be guilty.[47] Any criticism had to be muted and shared with only the most trusted comrades. Once several friends and former Cheka associates from Minsk visited the Alferovs in Siberia. Sitting around a table, eating and drinking, they shared stories about the extensive arrests in Belarus. They spoke in hushed and unusually critical tones about Stalin. Marx and Zhores, who were hiding on the steps nearby, overheard the adults and knew that they could never repeat a word.

In 1939 the western Siberian region was administratively divided into the province of Novosibirsk and the Altai Mountains region, and the trust moved its operations to Barnaul, further south on the Ob River. Since Ivan Alferov had reached a very high rank in the Commissariat of the Means of Communication, officials ordered that his family be moved by train in a wondrous furnished coach; in the absence of an apartment, they lived in the wagon on a railroad siding. During the summer they had a cabin on the Ob shoreline. Eventually the family got a three-room apartment in a newly built two-story wood home with central heating. The boys studied at the railroad school, about a ten-minute walk from the car.

Alferov, Literature, and Industrialization

On a business trip to Moscow, Ivan Karpovich Alferov met with the commissar of the recently formed Pulp and Paper Industry ministry, N. N. Chebotarev, another old acquaintance, who offered him directorship of the Siask Pulp and Paper mill near Leningrad.[48] Once again we must recall that the Alferovs were "typical" in the unexpected impacts of the revolution on their lives, but atypical in their family connections and unceasing upward mobility. Alferov gladly accepted the new post at Siask. When they left Barnaul just a few months later, in August of 1939, Zhores's parents decided to leave their dog Kabek behind. Zhores considered this a criminal act and cried without respite. They traveled back to Leningrad not in a first-class coach but in a typical car. Zhores's mother had grown accustomed to good service on trains and was shocked by the fall in its quality. She quickly figured out why and asked Ivan to wear his commissariat-issue railroad blazer with three stars on it so that the conductors would notice him. For the rest of the journey, they were treated like communist royalty. On the way to Leningrad, they stopped in Moscow for three weeks, staying at the Metropol Hotel, with its tsarist-era overstuffed furniture, then proceeded to Leningrad on the "Red Arrow" (an express train reserved for important officials) in an international coach. Ivan's brother Valerian ("Uncle Valya") met them at the train station in Leningrad. Valerian had lived in Leningrad since 1910. He and his wife, Fenya, lived on Red Street at Labor Square with two daughters, Zhenya and Valya. The family spent a few glorious days sightseeing and visiting the Hermitage and the Russian Museum. Zhores says he will never forget the view from the Palace Square, the Winter Palace, the Triumphal Arch, and the General Staff building.

Siasstroi was a factory town in the Volkhovsk region of Leningrad province about a hundred kilometers from Leningrad at the mouth of the Sias River, and not far from Volkhovsk, the site of the first Soviet hydroelectric power station built as part of the State Electrification Plan on the Volkhov River. Beginning in the early nineteenth century, tsarist engineers tried to improve navigation on the river. The Bolsheviks' determination to create communism through electrification led to the sponsorship of precisely such river improvement projects as dams, locks, and power stations. The Volkhovsk power station opened on October 19, 1926, and Soviet engineers proudly called it the "first born" hydroelectric project in the USSR. Engineers and Bolsheviks alike, including such technology enthusiasts as the tsarist-era electrical engineer Count Heinrich Graftio (designer of the Volkhovsk station) and Lenin himself, embraced the station as a trigger to further hydro-engineering the likes of which the world had never seen. It was a symbol of the rebuilding the Russian electrification system after the devastation of World War I, the revolution, and the civil war. The scale and nature of construction—reliance on a huge workforce of 12,000 laborers equipped with rudimentary tools—presaged the scale and the technological level of construction at huge Stalin-era power stations, which inspired Zhores Alferov to turn to electronics. In 1928 the hydropower station supplied more than 60 percent of Leningrad's power, and it operates to this day. At Volkhovsk the Bolsheviks built an aluminum factory (which spit out produced its first ingots in May of 1932 in celebration of the five-year plan) and the nearby Siask Pulp and Paper Combine, where Ivan Karpovich Alferov now worked.

The Siask Pulp and Paper Combine was another of the hundreds of new enterprises founded during the first five-year plan and intended to demonstrate the advantages of Soviet-style central planning over capitalism. The construction of what was at the time the largest mill in the nation inspired the novelist Leonid Leonov to write *Sot'* (often translated as *Russian River*[49]). Leonov was another connection between Zhores Alferov, pulp and paper mills, and rivers, both because Zhores's father was centrally involved in its operation and because Zhores took inspiration for his devotion to socialist causes from Leonov and from other lesser authors of the genre. In writing *Sot'*, Leonov drew on firsthand knowledge of Arkhangelsk. His father, a poet and a journalist, had been arrested for anti-tsarist activities and later exiled to Arkhangelsk, where he had set up a newspaper. Leonid Leonov worked for the newspaper, then turned to fiction.[50]

Sot' was not quite the typical Soviet production novel of the Socialist Realist genre. In Socialist Realism, artists, authors, playwrights, and film directors portrayed a glorious future in stark terms of right versus wrong, ally versus enemy, proletarian versus capitalist, Party activist versus "wrecker," and always in terms of final success. Leonid Leonov developed sophisticated characters and plots that revealed the challenges facing socialist society. *Sot'* focused on the obstacles encountered and eventually overcome in the building of a paper mill. These included natural disasters, opposition from allegedly hostile well-to-do kulaks (peasants) and from monks, ignorance, encounters with saboteurs and ill-wishers, and the pitfalls of building a new technology in a backward country. Like many other Soviet authors, Leonov was encouraged to visit the sites of "hero projects" to find inspiration for his work. He visited the paper mills of Siasstroi and the Balakhna Pulp and Paper mill on the shores of the Volga in 1928. He observed the dangers to workers, yet had enthusiasm for these factories, which were a metaphor "for the construction of a new society."[51] In short, he captured the challenges that socialist construction posed to such managers as Ivan Karpovich Alferov.

Sot' described many but not all of the problems Ivan Karpovich Alferov encountered in embarking on a career in the pulp and paper industry, including technological backwardness and inexperienced workers. In many of the facilities built and rebuilt during the industrialization effort, the authorities appointed a "red director" and a "bourgeois director." The bourgeois director was often a specialist who had been trained during the tsarist era or who had suspect political leanings that reflected his class. His knowledge was required to ensure the factory operated according to specifications. The red director learned how technology operated on the fly, in part by peering over the bourgeois director's shoulder. The workers had rudimentary knowledge of machines. Since Ivan Karpovich Alferov had long worked in mills and had graduated from the Industrial Academy, he did not require a bourgeois director. The director's cottage in Siask, nonetheless, was quite bourgeois, its five rooms filled with lovely pre-revolutionary furniture. The heating was by wood stove. In the winter of 1939–40, when temperatures dropped to –50°F, they could barely heat the house to +50°F. It was dark everywhere, as there was little electricity.

During the Russo-Finnish war, in the winter of 1939–40, the Alferovs bivouacked several soldiers on their way to the Finnish front; after the war, several of them stayed with the family again. (The Alferovs eventually settled in a house in Komarovo on the Gulf of Finland, land that the Finns would cede to the USSR after the war.)

Zhores continued his elementary schooling during this time, further developing his oratory skills and his modest knowledge of literature. Like all schools, Zhores's had a Children's House of Culture that promoted student interests through a series of "circles." His mother urged him to join the musical circle and study the piano. Eventually he gave in, though he claimed to lack musical ability. He quickly memorized a series of pieces, though he had no real understanding of music. When his piano teacher fell ill for a long time, he escaped that circle for the chess circle. He and Marx—and many other school-age children—were thrilled that Soviet grandmasters were the best in the world. Zhores especially loved Mikhail Botvinnik, the six-time Soviet and five-time world champion, and also followed Isaac Boleslavskii, Andrei Liliental, Vasilii Smyslov, and Paul Keres. Zhores himself showed skill at chess, but dropped it for academic interests.

Zhores's favorite circle was the public speaking circle, which led him and other children to perform at various concerts and in various venues, including at the paper mill. He recalls his repertoire to this day: Vladimir Mayakovsky's "Poems about a Soviet Passport," "Jaures," and "At the Top of My Lungs," and three stories of Mikhail Zoshchenko—"Aristocrat," "Krazha," and "Akter." He can still recite the poems of Mayakovsky, a leading poet of the early Soviet period and a founder of the Russian Futurist movement. Born in Georgia, Mayakovsky moved to Moscow early in life and joined the Bolsheviks. In 1924 he composed an elegy on the death of Lenin that made him a household name. He traveled to the United States, stopping in New York, Cleveland, and Chicago, and wrote poems about the Brooklyn Bridge and other symbols of American society.[52] In "At the Top of My Lungs" (1930), Mayakovsky addressed the liberation of the underclass and criticized intellectuals for analyzing but not acting on their armchair analyses. In "My Soviet Passport"[53] (1929), referring to customs and other officials, Mayakovsky acknowledged that citizens of other countries usually drew barely a notice, and often polite obsequiousness. But the holder of a Soviet passport created violent consternation, which Mayakovsky proudly endured when abroad. Zhores wrote school compositions on Mayakovsky's poems, and performed them to audiences at Volkhovsk, Svir, and other "hero" construction sites.

Zhores enjoyed reading Zoshchenko from the first time he read his satirical stories at the age of six or seven years. The early stories of Mikhail Zoshchenko dealt with his war experiences. He then turned to satire in a series of humorous short stories that highlighted the joys and sorrows of the increasingly corrupt and bureaucratic society and how people dealt with such daily problems as poor housing and food shortages. Zhores, with his sense of humor and Soviet propriety, finds the stories captivating to this day. As a small boy, he often retold Zoshchenko's stories to large audiences. One of his favorite was "Aristocrat," in which an apartment superintendent falls for a woman resident because of her stockings and gold tooth (she must be an aristocrat!), goes for walks with her, eventually takes her to the theater, but cannot afford to treat her to the pastries she gluttonously devours at the buffet during the intermission, and they break up over the incident. Unfortunately, in the 1930s Party hacks in the Writers' Union pressured Zoshchenko to conform to Socialist Realism, and he ceased writing satires.

Science and Bolshevism[54]

On the eve of the revolution, Russia lagged behind Germany, England, France, the United States, and other scientific powers in funding and other quantitative indicators of science. But by the time Zhores Alferov entered his middle school years, the Soviet scientific establishment had grown quantitatively and qualitatively, with representatives from population genetics, solid state and nuclear physics, physiology, and other fields recognized by foreign colleagues for their contributions. The revolution had left scientists vulnerable to the same political and economic uncertainties that buffeted society. With the provisional government unable to support them at the level needed, universities and research institutes could not buy subscriptions to journals, purchase reagents, or replace equipment, let alone heat buildings. The Bolsheviks promised support for science. Even if many scientists disliked the Bolsheviks, they welcomed support of the scientific enterprise.

Bolsheviks and scientists soon reached a mutually beneficial accommodation. For the Bolsheviks, especially under Lenin's influence, science was crucial to the future of the Soviet state. Members of one communist group, the Proletkultists, held the notion that a unique "proletarian culture" that would form after the revolution would enable the rebuilding of capitalist legal, economic, social, and cultural institutions. The Proletkultists set out to establish a proletarian university, to compile a proletarian encyclopedia, to write and perform proletarian literature, poetry, and music (some of which involved performers sounding factory whistles and using hammers to bang on metal), and to develop a proletarian science with its own epistemology and methodology.

Lenin and other more pragmatic, less utopian communists believed that an alliance between scientists and the new regime was both possible and necessary—for some indeterminate time—to rebuild the nation and its industry. For Lenin, scientists naturally had a materialist philosophy and worldview that fit the agenda of the Bolsheviks. Indeed, they believed that Marxism itself was a science, and that capitalist science, also a product of industrial Europe, could join Marxism in the construction of communism. As soon the Bolsheviks seized power, even though they faced opposition from all sides, and even though civil war broke out, they established several bureaucracies to support the growth of the scientific enterprise and to tie research more directly to the economy than had been the case in the tsarist era. Science and technology would create the modern industrial manufactory, electrify rural regions in support of agriculture, and transform transport and communications, with workers as the chief beneficiaries. On many levels, the slogan "Communism equals Soviet power plus electrification of the entire country" indicates that the Bolsheviks saw science as a panacea for their economic, political, and cultural programs. Zhores Alferov embraced Leninist views of science, especially endorsing the efforts to create new institutions of science and education.

In creating institutions to promote science and incorporate its achievements into the economy, the Bolsheviks were ahead of most countries, in particular the United States. Granted, the American system of universities and industrial laboratories (such as those of Westinghouse, General Electric, and DuPont) was the envy of many nations.[55] Yet in the United States federal support for research and development, as opposed to support of industry, took off much later, with the pursuit of nuclear bombs and radar during World War II, and not until 1951 did the government establish the National Science Foundation. The Soviet government therefore was a pioneer in government support of science. Of course, the Germans had already established major industrial laboratories, in particular in the chemical industry, and the institutes of its multi-profile Kaiser Wilhelm Gesellschaft, founded in the early 1900s, ensured a dynamic relationship between science, engineering, and the economy. However, hyper-inflation and the inattention of policy makers hurt those laboratories, the society, and the universities.[56]

Early Bolshevik leaders, many of them European-trained intellectuals, pushed by scientists, formed two organizations that explicitly encouraged the expansion of the scientific enterprise: Glavnauka and NTO. Glavnauka—the Main Scientific Administration—fell under the Narkomprosvet (Commissariat of Enlightenment or Higher Education). The commissar of enlightenment, Anatolii Lunacharsky, a longtime associate of Lenin and (like many of the first Bolshevik administrators) a well-educated man with years of experience living in exile in Europe, had a cosmopolitan view of scientists and their potential contribution to science and the Bolshevik future. He supported the expansion of Glavnauka's budget and bureaucracy to create a series of new research institutes. He believed that scientists ought to have relative autonomy in setting the research agenda, hiring employees, and traveling abroad to meetings and for study.[57]

Zhores Alferov's eventual home, the Leningrad Physical Technical Institute, was founded in 1918 under the jurisdiction of Glavnauka. (In the pages that follow I shall refer to this institute as the LFTI. It was also known as "fiztekh.") The LFTI was known as the "cradle" of Soviet physics, since through its doors passed dozens of future academicians and corresponding members of the Academy of Sciences and four Nobel laureates (Alferov, Petr Kapitsa, Lev Landau, and Nikolai Semenov). On the basis of its personnel and its equipment, a dozen or so other major physics institutes were founded through the USSR, as far away from Leningrad as Sverdlovsk in the Urals and Kharkiv in Ukraine. Its founder and director until 1950, Abram Fedorovich Ioffe (1880–1960), took advantage of Anatolii Lunacharsky's support and openness to build a formidable research program in solid state physics. Other scientists gained funding from Glavnauka to build other institutes and found scores of scientific societies, many of which maintained their autonomy into the late 1920s.

NTO, the Scientific Technical Department (later "Administration") of the Supreme Economic Council, set up at the urging of Lenin's personal secretary, N. P. Gorbunov, navigated through the ebbs and flows of Soviet political currents. In the mid 1920s, during the New Economic Policy, the economy began to grow, and with it the scientific establishment. Gorbunov had lofty goals for NTO to contribute to economic growth. He saw NTO as fostering the creation of a network of factory laboratories to advance modern production processes and promote innovation and invention; Bolshevik leaders believed that somehow, under socialism, science and innovation would be more closely linked than under capitalism, and the proletariat would experience the benefits in higher quality of life and work almost seamlessly. Gorbunov invited scientific experts to participate in economic and scientific planning. He complained to Lenin that creating NTO challenged him at each step. He believed it was necessary to destroy old forms of administration and create new ones, and that the paucity of resources—including communist personnel—made this a challenge. Under NTO, officials set up such new research centers as the Automobile

Laboratory, the Central Aerodynamic Institute, the Central Scientific Technical Laboratory, and the State Pulp and Paper Experimental Station.

After Gorbunov moved on to other work, NTO's fortunes were less clear. Such leading scientists and engineers as V. N. Ipatieff (a chemist of international reputation who eventually emigrated to the United States) and Peter Palchinsky (whom Lenin directly suggested serve as the head of NTO) were involved in its activities but complained about their powerlessness to accomplish much, in view of the highly bureaucratized nature of Soviet society and the mistrust of Bolshevik administrators they engendered. Several of NTO's communist directors—Leon Trotsky, Lev Kamenev, Nikolai Bukharin—were appointed after being removed from Party leadership posts. Each of these three men was occupied with a struggle against Stalin for political and personal survival that distracted him from science, and each was disappointed with his demotion to NTO. Bukharin sincerely attempted to improve the functions of NTO, but faced constant criticism for allegedly failing to take note of the class content of science.[58] While NTO therefore helped to establish and finance several important research institutes with close ties to industry, its functions, like those of Glavnauka, were eventually folded into such massive industrial ministries as Narkomtiazhprom (the Commissariat of Heavy Industry) in 1931 as part of Stalin's industrialization campaign. The new jurisdictional relationships left no doubt that science would serve the Soviet state and its economic programs. Still, in 1926 Ioffe was able to convince NTO to fund the Leningrad Physical Technical Laboratory, whose work was closely connected to factory laboratories. In fact, LFTL personnel, buildings, and equipment overlapped with—and were often the same as—those of the LFTI.

Abram Ioffe's ability to gain funding from Glavnauka and NTO for his institute may be the most prominent example of how scientists skillfully navigated among government bureaucracies to preserve some autonomy in the face of Bolshevik efforts to tie their research to the engine of "socialist reconstruction." Many scientists and engineers hated the Bolsheviks, who they saw as crude—and, they hoped, temporary—usurpers of legal power. As representatives of the upper and middle classes, they rejected the Marxist ideology of class war and notions of a proletarian science as distinct from bourgeois or capitalist science. They faced hostility from the workers of Petrograd and Moscow and mistrust from political administrators, as well as spontaneous expropriation of their property and homes. In the first years of the revolution, millions of people emigrated over fluid borders, many of them representatives of the artistic and scientific intelligentsia. It would become necessary to create a new Soviet intelligentsia, of which Ivan Alferov was a member of the first generation and Zhores Alferov a member of the second generation. During the shortfalls of food and fuel that hit Russia during the civil war, famine and epidemics also hit the nation, and scientists and other members of the intelligentsia may have suffered disproportionately. Seven of the 44 full members of the Imperial Academy of Sciences (15 percent) died during the civil war. The critical state of the intelligentsia led the playwright Maxim Gorky, a longtime associate of Lenin and Lunacharsky, to appeal to Lenin to create a special system for the intelligentsia of food rations, access to health care facilities, and the award of gold rubles, under the Central Committee for the Improvement of the Living Conditions of Scholars (TsKUBU), to save them. Gorky, a champion of the idea that Russia needed modern Western science to raise the level of culture of the backward peasant to the level of Europe, joined leading scientists in an organization for the advancement of science toward that end.[59]

Yet for many scientists, even those who mistrusted Bolshevik political motives, the new regime was a welcome change from the tsarist regime. Tsarist bureaucrats—and the tsars themselves—had mistrusted scientists, especially natural scientists whom they considered too liberal. Even in the era of the Great Reforms after the 1861 emancipation of the serfs, the Ministry of Higher Education closely controlled professional appointments, curricula, and student admissions. Granted, the tsarist government supported science before the revolution, though not to the level which scientists (especially those in the Academy) desired. Also, university education expanded after the emancipation. And such scientists as Dmitrii Mendeleev earned bonuses for consulting with the industry or with the Ministry of Finance—especially under Count Sergei Witte, who supported the expansion of polytechnical education in Odessa, in Petersburg, and elsewhere. Yet during the last three decades of Romanov rule, tension between the government and university remained very high; in 1911, more than one-third of the professors of Moscow University resigned to protest the policies of Minister L. A. Kasso, who seemed to be allied with the anti-Semitic Black Hundreds. The Imperial Academy of Sciences was dominated by conservative interests and by such disciplines as philology, literature, religion, and classical studies. The government funded research in the physical sciences poorly. Even when faced with a desperate crisis during World War I, the government only belatedly recognized the need to support research on explosives, weaponry, or resource use in a newly established council. The council itself was established only in 1916, two years after the war had cut Russia off from European resources and science. The Bolshevik interest in science expressed in Glavnauka and NTO, and in funding, was therefore a profound change.

Abram Fedorovich Ioffe, the founder of the LFTI in 1918 and its director until his ignominious removal in 1950, led the successful effort of scholars to rejuvenate science under Bolshevik power. (The LFTI remained the second-largest institute of the Russian Academy of Sciences in 2009.) He worked closely with fellow scientists and with Bolshevik administrators to acquire funding, to re-equip institutes, and to reestablish foreign contacts, including journal subscriptions. He secured a building, in the northern suburbs of Petrograd, whose previous owner had been displaced by the revolution; the building had become property of the Bolshevik state. This location ensured access to promising students, and a quiet environment 12 kilometers from the dirt and political turmoil of the center of the city.

Ioffe pursued a four-part program that transformed the LFTI into the cradle of Soviet physics. First, he selected a broad-based program in solid state physics to serve as the foundation of institutional research that grew out of his own 20-year studies of crystal structures using x rays. Solid state physics would also be Alferov's focus. The institute included or later created departments or programs in heat engineering, theoretical physics, and nuclear physics. Second, Ioffe worked closely with such higher educational institutes as the Polytechnical Institute to create new departments to train students for advanced research at the LFTI; the most important of these was the Physical Mechanical Department at the Polytechnical Institute, which is still extant. Alferov would also be heavily involved from his home institute, the LFTI, in promoting higher educational training programs at other institutes to ensure a flow of talented researchers. These programs— and the growth of the physics discipline generally—facilitated the postwar expansion of the LFTI. Third, Ioffe participated in a national effort involving other fields of science to resurrect scientific publications. Among other journals, the pre-revolutionary *Journal of the Russian Physical Chemical Society* (*Zhurnal Russkogo Fiziko-khimicheskogo Obshestva*, published until 1931) was restored, and the *Journal of Technical Physics* (*Zhurnal Tekhnicheskoi Fiziki*) and the *Journal of Applied Physics* (*Zhurnal Prikladnoi Fiziki*) were established. Other journals followed, including *Physics of the Solid State* (*Fizika Tverdogo Tela*, 1959–present), which Alferov edited for many years. These journals helped to demonstrate Soviet scientific priority in a variety of fields. (In the 1930s, to ensure readership of their research abroad, Soviet physicists at the Kharkiv Physical Technical Institute—which was founded as a spin-off of the LFTI—published a German-language journal, *Zeitschrift für Physik der Sowjet Union*, but the journal was shut down in 1937 as the

Soviet Union became more isolated.[60]) Fourth, and crucially, Ioffe led the effort to reestablish contacts with Europe. War and revolution had isolated Russian scholars. Their subscriptions to leading German, French, and British journals had been interrupted. They needed modern equipment. On several occasions, Ioffe secured gold rubles and headed small delegations of scientists to the West to buy instruments, reagents, and subscriptions. Soon Soviet scientists began to publish widely in such leading journals as the German *Zeitschrift für Physik*. They secured fellowships from the prestigious Rockefeller International Education Board and other foundations to study with Werner Heisenberg in Leipzig, with Niels Bohr in Copenhagen, in the United States, and elsewhere.[61] This dynamic environment, in which the scientific enterprise grew and Soviet physicists considered themselves members of the international scientific community, persisted until the rise of Stalin as dictator in 1929–30. Alferov joined the LFTI in 1953, and throughout his career, as an internationalist in science, he sought to expand and strengthen ties with the world scientific community against xenophobic and ultimately autarkic Soviet leadership.

To the Urals

During Zhores Alferov's childhood, his family entered the class of the new Soviet intelligentsia. The process of vydvizhenie advanced thousands of dedicated and often militant communists into positions of responsibility in various economic ministries, educational institutions, and cultural organizations. At the same time, the Communist Party employed coercion, violence, cooptation, and persuasion against members of the old intelligentsia, people who had received training during the tsarist era, and people who were members of the bourgeoisie or whose parents were. The system of dual management—having a red director stand at the shoulder of the old expert—was just one manifestation of the cultural revolution intended to create a new intelligentsia and ensure the allegiance of the old intelligentsia. Yet the violence against the old intelligentsia, and against alleged "wreckers" and others caught up in the intrigues of Stalinist politics, destroyed the careers and lives of many experts who might have contributed to "socialist reconstruction."

Out of sincere curiosity, and recognizing the role the family must play in this new world of education, science, and cultural revolution, the Alferovs actively promoted reading, writing, and scholarship inside and outside the home. Anna Vladimirovna worked as a volunteer in the library of the combine's House of Culture. Precisely at this time, the family began to acquire more and more books. Anna and Ivan realized that their peripatetic life had ended. Leon Feuchtwanger's *Moscow 1937* and Matvei Bronshtein's *Solar Matter* (the latter a book on helium and the universe for children) suddenly were forbidden.[62] Both *Moscow 1937* and *Solar Matter* should have been destroyed at the library by order of Communist officials, but Anna brought them home.

Bronshtein's *Solar Matter* (*Solnechnoe Vechshestvo*), published in 1936, captivated Zhores. One day I entered his office to find Alferov happily leafing through a first edition, one of a few old books that had turned up when he had unpacked several long-forgotten boxes. At the time, most people considered it the most successful children's book of the year if not the decade. "You can read this book at ten or eighty years old," Alferov told me. He also had Bronshtein's *Atomy i Elektrony* (1935) on his desk, a gift from his friend Viktor Frenkel', a historian of physics who saw to it that Bronshtein's books were re-published nearly 40 years after Bronshtein was executed and his books ordered out of circulation.[63]

Bronshtein, a theoretician, a specialist in astrophysics, and a pioneer of quantum gravity, was one of the shining stars of the LFTI. Like George Gamow (a Leningrad physicist who escaped the USSR in 1933 and wrote *Mr. Tompkins in Wonderland* and *Mr. Tompkins Explores the Atom* in the United States), Bronstein wrote popular science books for children. He was married to Lydia Chukovskaia, a writer and later prominent human rights activist and a friend of Andrei Sakharov. In 1937, during the Great Terror, Bronshtein was arrested and shot; his wife was told that he had been sentenced to ten years in the labor camps.[64] The Soviet astrophysics community—centered at the Pulkovo Observatory at Gatchina, outside Leningrad—suffered great losses. Perhaps 10 percent of Soviet physicists was arrested during the purges. Indeed, entire scientific communities faced purges in Leningrad, Moscow, and Kharkiv.

Many years later, Alferov, already a Nobel laureate, after mentioning *Solar Matter* in a television interview, received a lovely note from the daughter of Bronshtein's wife thanking him for recalling Bronstein. Had Bronstein lived, Alferov and others believe, he would have been another figure on the order of George Gamow or Lev Landau, perhaps another Soviet Nobel laureate.

One beautiful day in 1941, the minister of the pulp and paper industry, N. N. Chebotarev, called Ivan Karpovich Alferov to Moscow for consultation over growing intrigues at the paper mill. It seemed that one Butriashkin, the chairman of the Party organization of the mill, was doing all he could to become its director. Chebotarev, who knew Ivan well, suggested that Ivan and Butriashkin be put in charge of two different mills, perhaps as a trial run to see which was the better manager. The ministry had opened five pulp and paper mills since at the end of 1940, all connected with the gulag. This had been done in response to a letter that Georgii M. Orlov, an engineer at the Central Design Institute of the Paper Industry (Giprobum) in Leningrad, had written to Stalin. Orlov pointed out that another source of pulp might be needed in case of war, and that it should be produced from pine rather than cotton (which was much more expensive). Stalin ordered Beria to build five factories in short order, largely with forced labor, in different regions of the country: Perm, Sverdlovsk, Cheliabinsk, Arkhangelsk, and Turinsk. Orlov became a general of the NKVD and chief construction engineer for these projects. Butriashkin was sent to Factory Two and Alferov to Factory Three (in Turinsk, a small city near Sverdlovsk in the Urals). The Alferov family did not think this was a promising move, and for Marx and Zhores it meant the need to transfer to a new school. The move occurred on the eve of the outbreak of World War II.

2 Heroes and Hero Projects

Perhaps the most important event in Zhores Alferov's life was World War II. Young Zhores, his mother, and his father lived out the war in the Urals region, far from the front. They never doubted victory. But at what costs? During the "Great Fatherland Patriotic War" at least 20 million Soviet citizens died; several new studies raise the estimate of loss to 35 million. Millions more were maimed and lived out their lives their secluded from public view. The agricultural and industrial achievements of Stalin's five-year plans were destroyed. After the war, as a young student, Zhores Alferov applied himself to the rebuilding of the nation by continuing his studies. By that time Alferov had undoubted faith in the communist system and its ideals as he understood them. He believed firmly in the national leaders, and he embraced their commitment to create an industrial superpower on the foundations of modern science and technology.

For Alferov, the rebuilding and expansion of the system of higher educational and scientific research institutes was the greatest achievement of Soviet power. This system symbolized the essence of the communist promise to give every Soviet citizen the opportunity to find his or her true path. It signified the Party leaders' determination to defend the nation from future attack. (His later criticism of Presidents Boris Yeltsin and Vladimir Putin centered on their failure to give education and science their due support; see chapter 6.)

Because of his age, Alferov was denied what he considers the great honor of serving in the Red Army during World War II. He continued his education in junior and senior high school, although he was distracted from his search of a career path by the war, especially as he and his parents waited for letters from the front from the other son, Marx. He eventually selected physics as his specialty, went to university and graduate school in search of specialization, and entered the Leningrad Physical Technical Institute in 1953 as a junior scientific worker. There he established his reputation as a promising scholar in the area of the quantum physics of the solid state.

Between the day of the surprise invasion (June 22, 1941) and late 1944, much of the European USSR fell to German armies. The most productive farmland was ruined. The rural economy, which had suffered the terrible blows of collectivization and famine, now experienced armies that stripped villages of food, fodder, and equipment. Roughly 1,500 towns were leveled or burned to the ground. Many individuals important to the conduct of war—Party officials, scientists, engineers, and others—were evacuated hastily, just ahead of the advancing Wehrmacht. Workers loaded entire factories and research institutes onto trains for shipment to the east and the south, to Kazan, to Central Asia, to the Ural Mountains, and to Siberia. Two hundred enterprises were evacuated to the Cheliabinsk region alone, most between September and December of 1941. LFTI physicists, who on the day of the invasion had excitedly prepared to run the first experiments on a new particle accelerator, a cyclotron, saw the machine hurriedly disassembled and carted off to Moscow. A number of LFTI personnel were moved to Kazan, where they worked on demagnetization of ships for protection against mines. The nuclear program of Leningrad physicists did not recover until the mid 1950s.

Worse, Leningrad fell under a blockade. For 900 days, from September 8, 1941 until January 17, 1944, the city was virtually cut off from the outside world. Only an ice road over Lake Ladoga (the "Road of Life") managed to get supplies in and a few lucky individuals out. The ice road operated only in winter, and even then the trucks en route faced constant bombardment from the Luftwaffe. By the spring of 1942, a 1.5-kilogram loaf of bread cost 60 rubles, but could be purchased only on the black market. In desperation, people traded a gold medal earned in high school, a diamond ring, or a grand piano for a loaf. One-third of the population starved, froze to death, or died of disease. The survivors used sleds to carry the dead to the cemetery. Eventually they buried 690,000 Leningraders in vast, unmarked graves as big as football fields at the Piskarovskoe Cemetery.[1] Elsewhere, workers (including Valerian Alferov) toiled in cold factories with shattered windows and collapsed roofs.

In spite of his dictatorial rule and his failure to prevent war, Soviet citizens revered Stalin and believed he would save them from Nazi conquest. Yet Stalin had underestimated Hitler and ignored intelligence that indicated without a doubt the impending attack. When the attack came, Stalin disappeared for days in shock, requiring the Commissar of Foreign Affairs, Viacheslav Molotov, to address the nation about the Nazi invasion. In 1939, to postpone the war, Stalin had ordered Molotov to sign a secret non-aggression treaty with the German foreign minister, Joachim von

Ribbentrop. This had enabled the two authoritarian regimes to divide Eastern Europe, allowed Stalin to annex the Baltic nations with impunity, and permitted both countries to carry out wars of aggression—the USSR against Finland, Germany against Poland. Zhores Alferov recalls that many Soviet citizens viewed the Molotov-Ribbentrop pact as a temporary, tactical maneuver and expected war soon; many latter-day Russian historians consider the pact to have been a betrayal of the Motherland.

Stalin's disappearance at the start of the war shocked his closest advisors and contributed to the war's high human and social costs. When Stalin returned to lead the nation after ten days out of public view, he took control over war plans and rallied the nation to victory in spite of frequent tactical errors. In early July he finally appeared on radio to exhort the nation "All to the front!" He insisted on "No retreat!" He appealed to patriotism and to religious belief, even seeking the cooperation of the Orthodox Church and temporarily abandoning the effort to eradicate religion. Stalin, however, remained true to his paranoiac fear of enemies. By his Order 270, issued in August of 1941, a deserter's family was liable to arrest, and soldiers falling into captivity had betrayed the Motherland. Yet even in the darkest days of 1941 and 1942, when the Germans approached Moscow, blockaded Leningrad, bombed the Caucasus, and encircled Stalingrad, all Soviet citizens—the Alferovs among them—devoted their entire effort to securing victory against the hated "German fascist aggressors."

Ivan Alferov dedicated his service to the Motherland. His superiors ordered him to bring a pulp and paper factory in the Urals region up to full production and beyond—to Stalinist levels of output—in nearly impossible wartime conditions.

When the Soviets had finally turned the tide and were pushing Hitler's armies back toward Berlin, Marx Alferov died in one of the last tank battles of the war. Zhores lovingly remembers his older brother—a decorated hero—to this day. Some of the unassailable feelings he has about the glories of the Soviet system originated in deep patriotism that developed during what all Soviet citizens saw as a life-and-death struggle against fascism and in the sacrifice of his brother in that struggle. Zhores Alferov has studied the history of the war, both in homage to his brother's memory and because of his love of the Motherland. Alferov's patriotism and that of Soviet citizens generally comes as no surprise. The war experience contributed to the mentality of a state constantly at war against internal and external enemies that persisted until the collapse of the USSR. The loss of his brother, the postwar rebuilding effort, entry into higher education, the death of Stalin, the Cold War—all of these things shaped Alferov's views about the crucial place of education, science, and technology in protecting the Motherland.

From the Leningrad Region to the Urals

On the eve of the war, the Urals region, consisting of the Molotovsk (now Perm), Sverdlovsk, Cheliabinsk, and Chkalovsk regions and the Bashkir autonomous republic, extended more than 800,000 square kilometers and had more than 12 million inhabitants. It contained extensive mineral wealth—more than 60 different important elements and 12,000 cataloged sites of such strategic materials as bauxite, potassium, nickel, cobalt, titanium, tin, beryllium, bromine, magnesium, rubidium, cesium, chrome, vanadium, industrial diamonds, and copper, plus peat and lumber. For many of those materials, the Urals region was first or second in total reserves in the USSR, and first or second in extraction. It became the armed camp of the Soviet war effort, for the Soviets relocated much of their machinery and equipment and many of the people needed to operate it here to produce shells, bullets, tanks, transports, uniforms, helmets, and armor.[2]

In June of 1941, Marx Alferov had just finished the last day of school in Siasstroi, outside Leningrad.[3] Siasstroi, a company town with the largest pulp and paper mill in the region, was situated on the mouth of the Sias River, near Lake Ladoga. The morning of June 22 was sunny and warm, and there was a road race, in which Marx finished third. An orchestra on the balcony of House of Culture played military marches. Suddenly, an ominous announcement on the radio that Molotov would address the nation stilled the happy crowds. Zhores Alferov recalls that he ran up to the orchestra's conductor and tugged on his pants to get him to quiet the music. Molotov announced the surprise attack, the unprovoked perfidy of Germany, and rallied the citizenry to vigilant devotion to the task of destroying the invaders. He guaranteed victory.[4] Sviasstroi workers staggered home, dumbfounded. That evening, people sat around dinner tables, quietly yet heatedly debated the causes of the war, and assumed that the Soviet army was already approaching Warsaw, even as local air-raid sirens wailed. Ivan Karpovich came home with the news that the Alferovs would move immediately to Turinsk, a paper company town in the Urals region. The next day they traveled by train to Leningrad, where most people had already papered their windows to prevent shock waves from the shells from shattering them. After three nights with relatives, they traveled by train to Moscow. (Most people who wished to flee Leningrad could not buy train tickets, which were reserved for personnel deemed crucial to the war effort

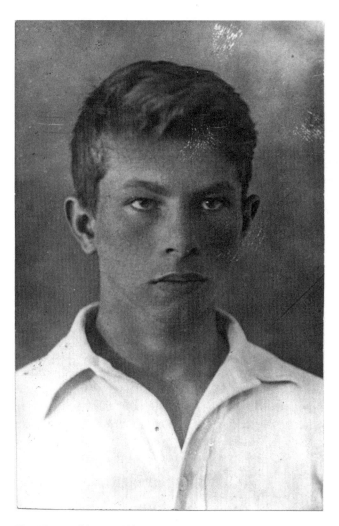

The 17-year-old Marx Alferov, 1941.

or the Party.) The fact that the Alferovs determined to leave Siasstroi and travel to Moscow even before national leaders acknowledged the invasion and declared war indicates how very well connected Ivan was to economic and political leaders. In Moscow the family stayed at the Metropol Hotel for a couple of days. The movies *Professor Mamlock* (which showed the evils of anti-Semitism) and *The Oppenheim Family* (about German persecution of Jews in the death camps) played in many of the capital's movie theaters.

The Alferovs traveled to Sverdlovsk in a heated step van with other evacuees, arriving in the early morning of July 3. They settled in the Bol'shoi Ural (Big Ural) Hotel, where they heard Stalin's belated war address on the radio. Stalin called on citizens to sacrifice everything to the war effort, to unite in the common goal of defending "every inch of Soviet soil," and to fight "to the last drop of blood for our towns and villages." Industry would be devoted to producing "more rifles, machine guns, artillery, bullets, shells, airplanes, and "we must organize the guarding of factories, power stations, telephonic and telegraphic communications and arrange effective air-raid precautions in all localities." Stalin called for a ruthless fight against "all deserters, panic-mongers, rumor-mongers; we must exterminate spies, diversionists and enemy parachutists, rendering rapid aid in all this to our destroyer battalions. All who by their panic-mongering and cowardice hinder the work of defense, no matter who they are, must be immediately haled before the military tribunal." Where retreat was required, everything that could not be saved had to be destroyed. Stalin concluded his address as follows: "All our forces for support of our heroic Red Army and our glorious Red Navy! All forces of the people—for the demolition of the enemy! Forward, to our victory!"[5]

With orders in hand, the Alferovs joined the war effort. They moved to Turinsk, 250 kilometers northeast of Sverdlovsk. At first they occupied two rooms in an apartment building. Ivan immediately set to work as director of a pulp and cellulose factory for the production of helmets, fuselages, gunpowder, and other war equipment. The family eventually moved to a house with an inadequate heating system; Zhores split wood constantly during the winter.

When the Alferovs and other evacuees arrived in Turinsk, the war had not yet had a significant impact on the region. Meat, honey, vegetables, butter, and potatoes were plentiful and relatively inexpensive. But in the autumn of 1941 the supplies disappeared, especially as new evacuees arrived, including others from Siasstroi.

Located between Solikamsk and Tiumen on the road to Tobolsk, Turinsk was a small town, although it was famous because several Decembrists, the first opponents of Tsarist autocracy, were exiled to it; in 1926 its population was only 4,500. In large part because of the influx of people and machines during the war, the population grew to 18,500 by the time of the 1959 census. The city and the surrounding region based their economy on the pulp and paper combine, a match factory, a motorcycle factory, and several logging companies. Agriculture also played a role in the local economy. The Turinsk Pulp and Paper Mill was founded in

1938 as Pulp Mill No. 3 of the NKVD to produce high-quality bleached cellulose.

Marx Alferov commenced classes at the Kirov Ural Industrial Institute in Sverdlovsk, intending to specialize in energy technologies. Marx was only 17 years old and thus not of draft age. Virtually all of the other institute students were women, since the older boys had gone off to war. Through the summer Marx dreamed of joining the army. In early September he informed his family that he had decided to volunteer for service in the Red Army. He hoped to go to the front; he simply could not in good conscience remain in the institute studying while the other students sacrificed their careers to the war effort. But because of his age the local draft board determined first to employ Marx, like thousands of other youths, in a factory geared to military production. They sent him to learn how to operate a lathe. Because of his leadership qualities, the factory Komsomol elected him secretary. Zhores Alferov feels it important to point out that his father's influence in the region was no less than that of the secretary of the district Party committee (Raikom) and no less than that of the chairman of the district Executive Committee (Raisispolkom). Only a telephone call was needed to keep his son out of the infantry or delay his service. But the thought never crossed the father's or the son's mind. Ivan had served at the front during the civil war, and the only thing that kept him out of this war was his responsibility to increase the production of cellulose as rapidly as possible. For Marx, youth delayed entry into the army until February of 1942, when he was sent to infantry school.

The winter of 1941–42 tested all of Turinsk's residents. The Alferovs had to trade many of their remaining possessions for potatoes. A good crop from their small garden in the late summer of 1942 made life somewhat easier, as did a special ration for such crucial personnel as factory directors, so they did not go wanting. Besides, the Urals were far from the invasion and were spared strafing by German airplanes. Families there donated money and clothing to the soldiers at the front.

The presence of hundreds of thousands of prisoners in the Urals region made the food situation tighter. The USSR employed millions of slave laborers in war tasks, toiling at the worst jobs in the worst conditions with the most rudimentary tools—and not only Soviet citizens but also prisoners of war—Germans Poles, Estonians, and Lithuanians, and other East Europeans, including hundreds of thousands of men and women exiled to the USSR well after the war's end. Zhores Alferov himself saw a large number of Germans arriving in Turinsk, and then also Tatars. Another large group of labor conscripts consisted of a million kulaks and their families who were exiled for their opposition to collectivization. Roughly 220,000 of them ended up in the region on the eve of World War II. Poles, Kalmyks, and Ukrainian nationalists were exiled in smaller numbers. The NKVD moved these groups around.[6]

Both Volga Germans and Crimean Tatars had been forcibly removed from their homes and their homelands on Stalin's orders. Party officials distrusted them and questioned their patriotism. By the eve of the Russian Revolution, Germans who had settled in the Volga region near Saratov in the eighteenth century had built a community of more than a million residents. Local Russians often commented on their economy, ingenuity, ability to work, and neatness. But the Bolsheviks hated the Germans for their Baptist and Lutheran beliefs and their language. One-third of them died during a famine in the early 1920s, and their clergymen constantly faced persecution.

On August 28, 1941, the government of the Soviet Union issued a Decree of Banishment that abolished the Germans' autonomous republic, and three days later they were forced to move north and east to the Urals. Young men were drafted into the Russian Army, and young women were used as domestic servants in the big cities. In 1944, the Crimean Tatars too were unjustly accused of being Nazi collaborators and deported en masse to Central Asia and other lands of the Soviet Union. Many died of disease and malnutrition on the way. Not until 1967 did the government remove charges against them, but no effort was made to return them to their homeland.[7]

Zhores Alferov saw both Germans and Tatars in Turinsk, as has been mentioned.[8] Many of them worked in the factory for his father. Zhores believes that they never complained, or at least never betrayed any anti-Soviet fervor, and were always friendly to the Russians around them. But these of course are the thoughts of a young man whose father was an authority figure to them. Why would they confide anything in him? In any event, he was surprised by the neatness and order of the Germans' barracks. While many of the factory laborers in Pump Mill No. 3 and other enterprises came from the Soviet countryside, Ivan also made use of "mobilized Germans," Tatars, and other gulag prisoners as employees.

In Turinsk, Zhores entered fifth grade. Many of his teachers had arrived from Kharkiv after the evacuation of the Machine-Building Technical School. Because of the need to place machinery and equipment from evacuated factories into any available building, the authorities requisitioned dormitories, theaters, stores, and even schools for production. The one remaining school in Turinsk grew so crowded that teachers had to work in three shifts. Zhores started his school day at 6 p.m., in the third shift. This left the daytime free for soccer, chess, reading, and study. He recalls that the teachers were excellent, especially those from Kharkiv. The mathematics instructor Aleksandra Ivanovna Malysheva, whose husband worked at the No. 3 mill, had received a "For Labor Valor" gold medal for her teaching. Zhores recalls her as proper and stern yet precise and clear. She so valued Zhores's study skills that she informed the students that in case of her illness he would conduct the lesson, and this actually happened on one occasion.

Physics and chemistry became Zhores's favorite courses. By the end of junior high school, he had put together his own laboratory at home with various reagents and apparatuses he had acquired from the paper factory. His attraction to physics extended to radio and electricity.

During the summers, Zhores behaved like any adventurous adolescent. In the summer of 1942, playing war games with friends, he stepped on a piece of broken glass and suffered cuts to a tendon and an artery. Blood came out in a fountain. Zhores ran to the nearby school to the nurse, but no tourniquet was applied. Fortunately the factory doctor was called and quickly took Zhores to a hospital. Zhores lost a lot of blood. The surgeon who stitched him up asked him, on the operating table, "Are you a Pioneer [that is, a member of the Communist organization for school children]?" In response to Zhores's "Yes," the doctor determined to complete the procedure without painkillers or anesthesia—drugs were in short supply owing to the demand from the front. The excruciating operation lasted 40 minutes, but Zhores says he did not cry. He spent part of the summer in bed and part of it hopping around one leg, but there were no complications.

The next summer, Zhores and several classmates went into the farmlands that adjoined the factory. Many factories in the USSR had their own fields, harvests from which served their workers directly through the dining hall, a snack bar, and home rations. The classmates were feasting on snap peas when a guard noticed them and fired a warning shot. The boys ran off, Zhores returning home with peapods filling the inside of his shirt. He entered the kitchen and was surprised to find his father at home much earlier than usual. When the boy tried to pass, Ivan grabbed his shirt and pulled it up, and out spilled the peas. For the only time in his life, his father struck him across the head. Zhores flew out of the kitchen, headed out the front door, and fell on the porch. His father rose and went back to work; Zhores sat sprawled against the wall and, sobbing, finished off the peas. Later Zhores learned that the guard had recognized him and called the factory's front office to report that the director's son was stealing peas. It was a good thing that Zhores's father was the director; this boyhood prank might have been considered a crime against the state.

Zhores Alferov hunting in Turinsk, 1944.

The Turinsk factory had operated at full capacity for much of the war, and any further increase in output was limited by the rudimentary technology. Nevertheless, in March or April of 1944 Ivan Alferov received a letter through the regular mail, not by secret courier, on the stationery of the Chairman of the State Committee of Defense:

Respected Comrade Alferov! I request that you increase production in the factory in May 30%. [signed] J. Stalin, Chairman, SCD[9]

Marx Alferov, War Hero[10]

At the end of February 1942, Marx Alferov, having volunteered for army service, was sent to infantry school in Sverdlovsk. In October he was sent to the front. Marx wrote his family frequently over the next three years. His mother kept the letters and gave them to Zhores, who has kept Marx's memory close to him by reading and re-reading the letters, who has published excerpts from many of them, and who sought context for them by reading extensively about the war. Anna Vladimirovna Alferov prepared for Marx's departure by sewing a stylish quilted jacket with a fur collar. Family members and close friends joined for a goodbye dinner, Anna Vladimirovna cried, and Zhores says that he and Marx cut up onions to hide their own tears.

According to Zhores, and as has already been noted, Marx had acquired great authority among the younger workers at the factory as Komsomol secretary, and among the older folk too. This authority carried over into the army, and he quickly became a platoon leader.

From the front, Marx wrote a series of letters that evoked the fever of battle, the depths of patriotic feelings among the young soldiers, and a sense that the tide had begun to turn by the spring of 1942. Marx had it easier than his married comrades who missed their wives, but "for me, beside you [family], I don't need anything." In another letter he assured his father that he would vanquish the enemy. He was ready to defend his homeland: "I'm going for our people, for their happiness and honor. I go for you, my family. In order to free Belarus where we all were born, and where all our relatives, my grandmothers, and where the gravestones of our ancestors are, where their homes have been burned down. I am going to destroy the aggressors at Leningrad." He instructed his father to get more out of the Turinsk factory, for the soldiers needed more gunpowder and bullets. By the winter of 1942–43, Marx was no longer a young man but a seasoned soldier who smoked and drank. And he proudly informed his parents that he had become a "candidate member" of the Communist Party. "Papa," Marx wrote, "now I am a candidate into that Party of which you have been a member since 1926." Marx could not wait to fight the Germans. "Today," he wrote to his parents in October 1942, "I am going to the front to defend the victorious socialist revolution from Hitlerism." Soon he would "meet the hated enemy with my own eyes." On October 26, 1942, he had arrived at the Stalingrad front as part of the Sixty-Fourth Army under General M. S. Shumilov. He knew the place. He had gone to school in Beketovka when the family had lived there, from 1935 to 1937, while he was in the fifth and sixth grades. After this battle, as Zhores discovered many years later in a letter from a military officer who had found the notice in an archive, Marx received a medal "For Valor" for his heroism in killing five German soldiers and, in spite of being wounded, leading his platoon to safety. Marx himself never told his family directly about his wounds.

The Battle of Stalingrad was the turning point of the war. Hitler was determined to make the city named after Stalin fall. Stalin was equally determined to save it. Hitler made the huge tactical error of trying to win control over the Caucasus and oil fields in the south while simultaneously attacking central Russia. This tactic left the German armies into a precarious position at Stalingrad with uncertain supply lines. In Stalingrad, Soviet and German soldiers fought fiercely for each neighborhood, each block, each building. The soldiers traded victories, day and night. They moved forward to claim some rubble, using hand grenades and rifles to advance, and were pushed back again and again. Marx fought bravely in the midst of the horror. His letters indicate that he managed to eat and to sleep, though not on a regular schedule of course. German bombers strafed the city, attempting to keep supplies open. But more and more the German soldiers were squeezed by the Soviet troops. Eventually, Soviet armies surrounded those of General Paulus, whose men lost their supply lines for armaments and food. Nearly 150,000 Germans died at the battle of Stalingrad (three times as many as the Russians claimed to lose), all Paulus's guns, motor vehicles, and other equipment were captured, and the Luftwaffe lost 500 transport aircraft.[11] The German Army, though far from being a broken force in early 1943, never recovered from the loss of an entire army, on top of millions of casualties already suffered on the eastern front.

As an avid amateur historian of World War II, Zhores Alferov has chronicled his brother's war experience against the background of Soviet history and literature, and has lectured on the subject of World War II. He especially enjoys the wartime correspondents of the novelists Vassilii Grossman and Ilya Ehrenburg. Alferov remembers well Grossman's vivid descriptions of World War II in the novel *Life and Fate*, particularly a terrifying scene in which Nazi soldiers execute the Jews in an entire village and his vivid descriptions of the Battle of Stalingrad; he considers *Life and Fate* perhaps the greatest novel of the twentieth century. Ehrenburg, however, is Alferov's favorite essayist of the war. He particularly loves *Buria* (*Storm*), published in 1948, which describes military actions in the Soviet Union and in France, the French resistance and the heroic efforts of one Madeau, the enormous efforts of the Red Army to defeat the Nazis, and the massacres of Jews at Babi Yar; even though it refers obliquely to the Hitler-Stalin pact, Ehrenburg received the Stalin Prize for it.[12]

To this day, Zhores Alferov expresses his strongest feelings of love and patriotism for the Soviet Motherland when thinking about Stalingrad and Marx's contribution to the war with his life. He rejects various revisionist anti-military tracts, published in the Russian press in the 1990s and the 2000s, that question Stalin's tactics, the nation's preparedness, or the ultimate morality of this war. He recalls instead a letter from his brother, dated February 6, 1943, that describes how Nikita Khrushchev, then first Party secretary of the Ukraine, met the tattered but victorious Soviet soldiers who had been gathered on the main square in Stalingrad to thank them for saving the city from the fascists. (Khrushchev presided over the purges of the Ukrainian Communist Party in 1938 and 1939 on Stalin's orders. He had brutally pushed the Ukrainian peasants to sacrifice again and again.) The main square hardly resembled a square; not one building in the city remained unscarred and most were rubble. The bodies of German soldiers still filled the streets and courtyards. The authorities let the soldiers rest for a few days, which meant time to bring order to the rubble and to dispose of the bodies. Marx and his division then were sent immediately to the front at Voronezh.

Voronezh tested the soldiers. Supply was miserable; they got 150 grams of flour a day and nothing more. By now Marx had earned a Red Star medal. Upon receiving the award, he remarked "It would be better had they given me a week of leave to home." The Red Star was an officer's order of the lowest grade, usually given for service, and during the war relatively commonplace, so it is difficult to imagine it could be traded for a few days of leave, so Marx's aside meant nothing. But these innocent words made their way to the political commissar of the regiment, who determined to punish Marx. The young soldier's personnel folder was brought to the attention of a meeting of the regimental Party organization. The Communist front-line soldiers, who knew from firsthand experience what Marx had meant and knew he was a true patriot, nonetheless spoke in support of the political commissar who demanded punishment. They were frightened of the consequences they might face if they were to remain silent or defend him. One regimental commander then spoke on Marx's behalf, praising him as an officer who loved his soldiers. He asked: Why should his personnel folder a matter of any significance in these circumstances? The Party organization fortunately and unexpectedly dropped its investigation.

In June of 1943, Marx Alferov awaited another battle as his regiment moved along the front toward Kursk and Ukraine. The troops marched 60 kilometers a day, even in the baking sun. It was logical for the Germans to move on from Kharkiv, which they had captured, toward Kursk. Hitler was determined to avenge his loss at Stalingrad. Stalin ordered Marshall

Zhukov to attack the Germans first. But Zhukov convinced Stalin to wait, and instead allow the Germans to wear themselves out and to trap themselves in the salient. He ordered the citizens in and around Kursk to build tank traps and minefields, while the army prepared anti-tank guns and dug in tanks to meet the German attack. The Red Army amassed 1.3 million men, 3,600 tanks, 20,000 artillery pieces, and 2,400 aircraft, the Germans 900,000 men, 2,700 tanks, and 2,000 aircraft. In the three-month battle, 100,000 Germans were killed or wounded. The Germans were forced to retreat and were on the run for the rest of the war. The Soviet government released figures of their losses only decades later: 250,000 killed and 600,000 wounded, and half of their tanks destroyed.

Marx Alferov was critically wounded on the third day of battle, July 7, by a mine that exploded a meter from him. Within a day only seven of the men in his platoon were alive. After the battle, a peasant saw him bleeding on the ground, noticed he was still alive, lifted him into a cart, and took him to a field medical station for an operation that saved his life, although he hovered near death for two weeks. On July 24 he was somehow evacuated to a hospital in Barnaul on the Ob River, where he recuperated. Hearing of his critical injuries, Anna Vladimirovna took leave from work to go after him and assist in his recovery. On October 10, mother and son returned to Sverdlovsk, where they were met by Ivan and Zhores, and the family was united for the last time. Three days later, Marx received orders to go to the front again, not to his own division, but as a reserve officer.

The family received Marx's last letter, dated January 6, 1944, with great delay, and then heard no news until May of 1944, when the dreaded official communiqué arrived stating that on February 15, 1944, Marx had died in battle. Ivan Karpovich was in Moscow when letter came. When he returned from his trip and walked in the door, he knew from Mama and Zhores's eyes what news had befallen them, and lay silently on his bed, crestfallen. Marx had died in the Korsun-Shevchenkovsk battle, in the Khil'ki Korsunsk region of Kiev province. He had survived the greatest battles of the war—Stalingrad and Kursk—but not this bloody skirmish. As is true of the majority of Soviet soldiers, no one knows how precisely how he was killed.

In 1956 Zhores decided to visit the village of Khil'ki with a close friend from the LFTI, Boris Petrovich Zakharchenia.[13] They took a train to Kiev and a boat to Kanev, hitched rides with peasants in trucks, then walked from village to village. In the center of the village of Khil'ki they found a memorial to Marx and the thousands of other heroes of battles in the region. As they stood in silence, a few older women approached them and

Zhores Alferov in 1949 at Komarovo, on the Gulf of Finland, where he and his wife, Tamara, would eventually settle. Komarovo has long been a second home to many of the Soviet intellectual, cultural, and political elite.

asked who they were. Zhores relates how they brought out tables, chairs, and tablecloths and laid on a feast. The grandmothers retold how the horrible battle forced them to hide in the cellars. Dozens of German tanks were destroyed in this village alone. In a field nearby, 5,000 Germans and 3,000 Soviet soldiers fell, including perhaps Marx.

Alferov again visited his brother's gravesite in 1983, this time accompanied by his wife, Tamara Georgievna, while on a business trip to the Ukrainian Academy of Sciences in Kiev. Upon hearing of his desire to visit Khil'ki, the president of the Ukrainian Academy of Sciences, Boris Paton, then a member of the central committee of the Communist Party of Ukraine, telephoned the secretary of the regional Party committee. The Academy provided a car and a driver, and the Alferovs arrived a few hours later. They were met by the second Party secretary in Khil'ki and the chairman of the agricultural council. It is hard to imagine such service had Alferov and Paton not been leading academicians and high Party members. The Alferovs visited a museum of the battle and saw a documentary film. Two years later, while in Volgograd (earlier "Stalingrad") on another business trip, Alferov met with the secretary of the Volgograd provincial Party committee. In a long discussion about business and science, the Stalingrad battle came up. The Party secretary, Vladimir Kalashnikov, upon hearing that Alferov's brother had fought there, called in an assistant, who located a museum associated with the 94th Division, in which Marx had fought. At the museum, Zhores discovered the address of the regiment commander, Mikhail Aglitskii, who now lived in Moscow. Returning to Moscow, Alferov looked up Aglitskii and called him to arrange a meeting. He bought along a bottle of cognac and some delicacies. They discussed the battle and life at the front generally. Alferov told Aglitskii that his brother had written in one of his last letters about his regret at not being able to return to his own platoon. Aglitskii remarked this had been all for the better, since after Kursk nothing remained of the group.

The Family Returns to Belarus[14]

Huge individual factories that dominated the landscape as kind of a Soviet company town were the distinguishing feature of the immense Soviet production enterprise. The factory towns were centered on local resources: production of iron and steel in Magnitogorsk in the south Urals region; nickel and copper in Monchegorsk; asbestos, concrete, and chemicals in other cities; paper, pulp, and lumber in Karelia, Arkhangelsk, western Siberia, Belarus, and other regions. After the war, the paper industries in Karelia and Belarus lay in ruins. Stalin determined that the resurrection of these industries in these towns was to be the first priority of the postwar five-year plan. The factories were usually up and running before housing, schools, hospitals, and shops were rebuilt. For Soviet citizens generally, rather than celebration, the war's end brought repression, re-establishment of strict Party control, and emphasis on reconstruction of the nation's industry first, not its housing stock, consumer goods, or other direct measures of the quality of daily life.

Toward the end of the war, Ivan Alferov was summoned to the Commissariat of the Pulp and Paper Industry in Moscow. The Commissar, Georgii Mikhailovich Orlov, was prepared to send Alferov to Enso (now Svetogorsk) on the Finnish border, the site of one of the largest paper mills in Russia, as the director of another combine. Orlov had Alferov's personnel file on his desk top, and all the paperwork was in order. But recalling that Alferov was Belarusian, Orlov decided instead to send him to Belarus to resurrect its paper and pulp industry. Alferov returned to Turinsk on May 9, 1945 (Victory Day), told Zhores and his mother about the new posting, and departed for Belarus. Zhores and his mother followed when the school year ended. En route to Belarus, they arrived in Moscow in mid June as the city prepared for a victory celebration and stayed on Arbat Street in the apartment of the director of the Main Administration of the Cellulose Industry, the Bychkovs. Alferov spent the next days with the Bychkovs' daughter Liuda, attending films, walking through the city, and relishing the victory. Citizens from all walks of life took part in a city-wide parade dedicated to the victory. In the parade, heavily decorated war heroes and soldiers led the way, followed by workers and then all others who wished to join in. Aside from the Harley-Davidson motorcycles (acquired through the American lend-lease program) that filled the city, Zhores's most vivid memory of the Victory Parade was that his father assumed an honored place on the visitors' tribunal in Red Square. Zhores Alferov's short stay at the Arbat Street apartment reveals once again how well placed his father was. Neither he nor his parents were elitist or class conscious. They were proud of their proletarian roots. But because of his father's prominent career in the pulp and paper industry, Zhores experienced a quality and texture of life unlike what ordinary citizens experienced. On the few occasions when he fell into trouble—stealing snap peas or hopping a train to travel somewhere without written permission—his father or someone who knew his father was able to assist him. Since the late 1920s, many members of the Communist Party and government elite and their families had lived on Arbat Street or in the neighborhood surrounding it. Many of the officials who were purged during the Great Terror disappeared from Arbat apartments.

Anatolii Rybakov's novel *Deti Arbata* (*Children of the Arbat*), which Zhores read after its publication in 1987, is one of the most interesting accounts of life on Arbat Street in the Stalin era. Rybakov finished *Children* in the 1960s but was not permitted to publish it until the Gorbachev era. The novel concerns a group of childhood friends from the Arbat district over a ten-year period beginning in 1933 (on the eve of the arrest of the principal character, Sasha Pankratov) and into the war years. For many readers the novel captures the events and tensions of the lives of members of the elite in the Stalin era better than many historical accounts. Alferov considers it superb precisely for the memories it inspires in him of his brief stay on Arbat street and its historical relevance.

From Moscow the Alferovs took a train to Minsk, the capital of Belarus. Arriving at the main station, they were shocked to find that virtually the entire city, including its beautiful decades-old train station, lay in rubble. They were met by a driver, who took them to their apartment through the ruins and past huge craters and burned out homes. A few large buildings remained. The Germans had planned to destroy the government, Central Committee, and Officers' Club buildings in front of the eyes of the liberating Red Army, but the Red Army liberated the city before the Germans could carry out that plan. Most people lived in the rubble, but the representative of the local paper factory had secured a small house on the Svisloch River for the Alferovs.

In the summers of 1945 and 1946, Ivan Alferov traveled frequently to various pulp and paper enterprises and lumber mills in the province. He often took Zhores along on these trips. In the absence of hotels, the two often spent nights with complete strangers. Zhores found the poverty oppressive. In the western parts of the province, the peasants had dirt floors, and wherever they stayed cockroaches and bedbugs overwhelmed the house.

The hot summer forced Zhores to learn to swim, an activity that became a lifelong love. As he tells it, on one occasion several senior officials arrived from the ministry in Moscow and joined his father Ivan for a dip in the Svisloch River. One of them called out for Zhores to join them. Ivan answered that Zhores did not swim. Zhores remembers that he was so embarrassed by the comment that he gathered all his strength and what little swimming knowledge he had, walked onto a bridge, and jumped into the water, then dog-paddled and flailed his way to the other side and back. After that, he spent the summer in the Svisloch River, swimming up and down, against and with the current. He has swum for exercise ever since.

First Studies in Physics and Mathematics

Owing to the disorder and devastation, and to the priority of rebuilding, few schools in Minsk had been able to open in the autumn of 1945, and of course they were poorly equipped. Only one boys' school and three girls' schools were open. Most boys were involved in the rebuilding effort or were in technical and vocational schools. At least Zhores Alferov was able to go to school; many other children had to wait weeks and months for schools to open.

Entering the ninth grade, Alferov began intensive studies of mathematics and physics. He also followed his healthy love of Russian and world literature, learned the Belarusian language, and read popular Belarusian authors. And he became acquainted with world literature, especially Shakespeare.

Alferov believes that his choice to study physics, his entry into a higher educational institute in Leningrad, and his becoming a scientist owe much to his excellent teachers—especially his physics teacher, Iakov Borisovich Meltserzon. In spite of war-damaged facilities, the students learned a great deal under Meltserzon. The school had a wood stove, and the students cut and split the wood. There was no physics laboratory. The key was that Meltserzon did not just read from the book but presented mature, engaging lectures and treated the students as adults. He refused to believe that anyone could find physics boring. He did not waste time asking students to solve problems in class, nor did he give quizzes; there were only quarterly exams. Zhores received a B in the first quarter, a B in the second, and a C+ in the third. He loved physics and deeply admired his teacher. Alferov believes that Meltserzon had marked him down for failing to use scientific notation. At the time, he thought the teacher was wrong for marking him down, and he was doubly hurt because he liked and respected him so much. Zhores's mother recognized his disappointment on his face. She attended the next parent-teacher conference and explained her concern over Zhores's lower-than-expected marks. In the next day's class, Meltserzon determined to make an example of Alferov for questioning his grades and called him to the blackboard for what turned out to be a lengthy interrogation. Meltserzon asked him to solve all the problem sets for the recent classes. Then he turned to the previous quarters, probing Alferov's command of that material. He continued the interrogation into the next class, again calling Alferov to the blackboard. He asked questions about ninth-grade physics, and then about eighth-grade physics. The exhausted Alferov's head was swimming by the time Meltserzon admitted that he knew his physics. He then changed the C+ to a B+. After that time, he would occasionally make a point in a lesson, explain it, and ask Alferov to comment. Alferov would stand, briefly discuss the problem, and get an A. Because of his mastery of the material, he had become Meltserzon's favorite student. In the tenth grade, Meltserzon, discussing radiolocation, explained how a cathode oscillograph worked. Alferov was enchanted with the device, electronics became his main interest, and within a few years was studying optoelectronics at the LFTI. As a college student, Alferov stayed in touch with his former teacher. Later still, when Meltserzon had retired and Alferov had joined the staff of the LFTI, they continued to meet from time to time.

Alferov had another excellent teacher in high school: Raisa Grigorievna Baram, who had won an award as Belarus's teacher of the year. He so appreciated Baram as a teacher and friend that later, when he visited his parents in Minsk, he always visited her.

The high school was at Freedom Square in what once had been the home of a tsarist governor. The students sat three to a table on uncomfortable chairs, Zhores Alferov with Igor Atrakhovich and Igor Aleksandrovskii. The three boys quickly became friends, and often they came to school together. Igor Aleksandrovskii was the son of the Belarusian soprano Larisa Pompeevna Aleksandrovna, whose roles included Lisa in Tchaikovsky's *Queen of Spades* and Margarita in Gounod's *Faust*. Igor Atrakhovich was the son of the Belarusian writer and dramatist K. K. Atrakhovich, who, under the pen name Kondrat Krapiva, wrote *A Ram with a Diploma* (1926) and *My Darling* (1945). A Communist Party favorite, Krapiva (1896–1991) won seven Orders of Lenin, an Order of the Great Patriotic War, a gold Hammer and Sickle medal, and a Hero of Socialist Labor award. One summer, Zhores sat in the library in the Krapivas' apartment and read through an eight-volume collected works of Shakespeare, developing an appreciation for Shakespeare that continues to the present.

At home the Alferovs spoke Russian, even though they had grown up in Belarus. But at the Krapiva house Zhores spoke Belarusian. Igor, his younger sister, and his parents were Zhores's true teachers of Belarusian. He frequently attended performances at the Academic Dramatic Theater of the plays of Kupala, Krapiva, Gorky, Simonov, and Ostrovskii, many performed in Belarusian, with Igor or his parents. He saw many of the leading actors and actresses of Belarus. This experience may explain his love for Belarusian society, language, and literature, which continues to this day.

Next door to the Alferovs lived Zhora Matsuk, two years older than Zhores, who had served in the partisans during the war and had acquired a German bayonet. Zhores coveted the bayonet. When playing with Matsuk he couldn't keep his eyes off it. Not surprisingly, Matsuk's school records disappeared during the war. After the war, he decided to enter a Sports Education School. A former partisan, he needed only fair grades on his entrance exams. There were three exams: written and oral exams on Russian literature and a written exam on the USSR's constitution. Though Matsuk could handle the written exams, he froze up during the oral exam. The examiners did not have photographs of the examinees, only a list of names. Matsuk, knowing of Zhores's lifelong love of literature and his knowledge of both the classics and contemporary authors, implored Zhores to take the oral exam for him. He eventually convinced him to take the exam by promising the coveted bayonet as an incentive. Zhores easily and quickly answered the questions about literature. Then he faced more probing questions. Indeed he faced surprise, because the examiners had noticed a significant improvement over Matsuk's performance on the written exams. Recognizing that his ease disturbed them, Zhores answered with greater uncertainty, pausing and stumbling from time to time but still achieving a passing grade. Matsuk gained entry to the Institute of Physical Culture, graduated with honors and later became the head of a department and a champion diver. Zhores only half jokingly considers the development of Soviet swimming and diving to be one of his accomplishments.

Since the school Alferov attended was not only the only all-boys school in Minsk but also the school for many children of the elite, leading politicians frequently dropped in deliver addresses there. Alferov believes that students' reactions influenced several politicians' careers. In October of 1945, for example, Mikhail Vasilievich Zimianin addressed the students. Zimianin, then first secretary of the central committee of the Belarusian Komsomol, would later serve in the diplomatic corps and, under Brezhnev, as a secretary of the Central Committee of the Communist Party in charge of ideology, censorship, and the press. At the school he spoke about the recent International Congress of Youth in London, which he had attended. Alferov recalls his speech as brilliant in content and in form, with a high degree of lyricism and wonderful humor.

In November of 1946, when Alferov was in tenth grade, Nikifor Iakovlevich Natalevich, chairman of the presidium of the Supreme Soviet of Belorussia and later a president of the republic, spoke about the fourth five-year plan to rebuild the economy. The school and the neighborhood crawled with secret police. Both the presence of so many police and the poor speech bothered Alferov and his classmates. Natalevich got lost in his numbers and couldn't answer questions very well. The students thought they knew more about the fourth five-year plan than he did.

The fourth five-year plan was a celebration of the reestablishment of the Stalinist economic and political system. Party leaders required the people to continue to pay for the war and to focus their energy on the reconstruction of heavy industry and the military, not pausing to reflect on their miserable living conditions. Party officials chose to rebuild the state and its apparatus, not to celebrate individual potentialities or reward workers and soldiers for the hard-fought victory. The plan again set forth superhuman goals for industry and the military, while health care, housing, and consumer goods lagged significantly in investment. Many people lived in shallow holes (zemliaki) or in tents and barracks, although for a select few in major cities Stalin ordered the building of huge apartment buildings, including seven "wedding cake" skyscrapers that arose on the Moscow skyline. These, too, were to signify the glory of the state. Agriculture continued to perform poorly; the consolidation of 250,000 collective farms into 90,000 megafarms was intended to overcome the lack of adequate investment and the absence of initiative among peasants, who hated the collective farms. But amalgamation had a negative impact on production. In 1946–47 a famine struck Ukraine, Moldova, and the lower and middle Volga region, and perhaps 2 million people died.[15]

Soviet scientists had also expected relaxation of ideological controls and the opportunity to become members of the international scientific community again. After all, Stalin had trebled their salaries after the war, the Allies had worked together in defeating Germany, and the 220th anniversary of the Academy of Sciences in the autumn of 1945 was an international affair, attended by large numbers of foreign scholars. Yet for scientists, too, the sense of relief was short-lived, with Party ideologues acting quickly to assert the priority of ideological purity in matters of science, politics, and philosophy.[16]

The tenth graders may not have been aware of the human costs of the plan, but they recognized Natalevich's superficial understanding of it, or at least they thought his presentation showed them little respect. As they left the auditorium, they saw some six- and seven-year-old schoolboys playing near the school's entryway in the first snow of the year. They advised the boys that a "really bad apple" would soon be leaving the school, and that they should be prepared to heave snowballs so that no dry spot remained. The little boys then showered Natalevich with snowballs, knocking his hat off. The police grabbed the offending children but really could not punish them. They decided to treat the matter as a rude childhood prank, and they whisked the minister away in a car. In December of 1947, the government ended the ration-card system and introduced a currency reform. A few days before the announcement of the reforms, Natalevich's wife had made several purchases at Minsk's only jewelry store[17]—purchases that were noted in the press. Natalevich lost his Party and government positions and was demoted to a paper-pushing job as head of an agricultural department in distant Penza.

The school's Komsomol organization, to which Alferov belonged, occasionally invited the secretary of the Ministry of Internal Affairs Komsomol organization to its meetings. The ministry's proximity to the school (it was just across the street) partially dictated this diplomacy, as did the school's leading position. On January 2, 1946, the police club in the Ministry building held a gala New Year's party. They put up a huge evergreen tree in the activities hall. In a neighboring room of the foyer the students painted scenes of fairy tales on cardboard boxes. They used cotton batting around the tree to mimic snow. They also erected a sixteen-foot-tall Grandfather Frost. Suddenly two students, perhaps fifteen-year-olds, ran from one of the side rooms. Going to investigate, Alferov and the others found the cotton and cardboard on fire. It quickly spread to Grandfather Frost. Though they were unable to put it out, it consumed the foyer and cut off the exit to those dancing in the activity hall. The boys led the girls out of the foyer and upstairs, but the fire continued to spread. Children jumped out of windows. About 100 people perished from the smoke and fire, and some were killed or injured jumping. Alferov and several other students did their best to help people escape the flame and gave first aid. Alferov told the secretary of the Komsomol organization that he had seen two youths run by him just before the fire broke out. Later the authorities identified the two. They were students from a trade school who had entered without tickets to the evening event. After smoking, they had thrown their still-smoldering cigarettes on the floor, the cotton had caught fire, and they—not having noticed, or perhaps having becoming quite frightened—had fled the scene. Both were punished with long prison terms.

Soviet justice was in any event stern. In the winter of 1946, in House of the Officers in Minsk, Soviet officials held a series of trials of German war criminals. No other republic—perhaps not even Ukraine—suffered such losses of life and personal property as did Belarus during the time of the German occupation. Somehow the Soviets held on to power in several regions of Belarus. Some kolkhozes (collective farms) managed to harvest grain and produce bread for the partisans in the autumn, with German soldiers afraid to go into those regions. Still, many Belarusians sided with the Germans and participated in the annihilation of the Jewish population. This was another tragedy for the Jews, who had faced pogroms, then purges in the 1930s. Partisan activity was more pronounced in Belarus than in other regions, and therefore Nazi reprisals against residents were brutal even by eastern front standards. Punitive operations by the Germans in January, February, and April of 1944 had left entire villages leveled, their inhabitants lined up and executed. All told, an estimated million people, including nearly the entire Jewish population (and many of the republic's most educated citizens), were exterminated in summary executions, in massacres, and in death camps in Belarus.

When German war criminals faced trial, Ivan Alferov and other officials were invited to attend the sessions. Ivan occasionally gave Zhores his ticket. Zhores remembers that one "Captain Koch," about to be hanged in a mass execution with sixteen other Nazi war criminals, admitted before the judge that he had personally killed 300 peasants, steadfastly contended that they were all bandits, and insisted that he regretted not having killed more of them. Perhaps this was Reichskommissar Erich Koch, who, as the Nazi official installed to exploit Ukraine in "colonial fashion," treated the Slavs as sub-humans intended only to serve the state.[18]

The Nazis intended not only to destroy Bolshevism, but then to occupy the eastern Slavic lands as Lebensraum (living space), with Jews, Gypsies, and others murdered and with Ukrainians, Russians, and Belarusian Slavs enslaved to perform agricultural tasks. The Nazis created civil administrative districts to carry out this murderous plan. In Ukraine nearly 4.1 million civilians died under Nazi rule. Of the 2.8 million laborers forcibly deported from the former Soviet territories to Germany, perhaps 2.3 million were from Ukraine. In Minsk, occupied without serious opposition shortly after the June 22, 1941, invasion, the Nazis established the Reich Commissariat Ostland under Wilhelm Kube. By late July of 1941, the Germans had established a ghetto in the northwestern part of Minsk. Perhaps 100,000 people were crowded into it to be used as slave laborers. Many of them eventually were murdered or died of disease or starvation.[19]

High School in the Postwar USSR

Beginning in the early 1930s, Soviet authorities established a rigorous system of higher education that, for students on the science track, emphasized rote learning of physics, chemistry, and mathematics, doused with the not-too-subtle flavorings of Party history and the Soviet epistemological system, dialectical materialism. Over the decades, the system trained hundreds of thousands of highly qualified specialists, but also, according to critics, led to arbitrary and needless checking on students' political reliability and thwarted the creativity of many of them. But in any system the best students succeed in demonstrating talent.

Zhores Alferov's exams in 1946 were a curious mix of testing in mathematics, physics, and literature as well as political acumen. Success on some of these exams might have been a matter of chance. Students drew a number from a list of perhaps 30 questions and had 45 minutes to prepare to answer one or two of them orally; if they had not prepared for a particular question, or could not show their knowledge of a subject while trying to answer it, they might receive a C or a D. Alferov received a B– on his Darwinism exam. Darwinism had come under assault from a number of biologists who embraced a Lamarckian view of the prevailing role of the environment and the inheritance of acquired characteristics as opposed to genes. Some of them believed in pseudo-cooperation, not a Darwinian struggle for existence among and even between species. They ultimately rejected modern genetics. The followers of this view were called Lysenkoists after their leader, the quack scientist Trofim Denisovich Lysenko. In 1948, with the direct intervention of Stalin, at a national conference called at the Lenin Academy of Agricultural Sciences in Moscow, Lysenko presided over the expulsion of geneticists and genetics from Soviet universities and research institutes, and even from textbooks.[20] In 1946, Alferov's grade was soon raised. After 1948, the outcome might not have been so favorable. Alferov, who occasionally glosses over such ideological intrusions of Stalinism into science, admitted "Of course, it's terrible when politics and power interfere in science anywhere, but especially during the Lysenkovshchina."[21]

In another episode, a large number of students got poor grades on an astronomy exam, perhaps because the teacher graded by alphabetical order, not achievement. When the students complained, the teacher was sacked from that job and assigned to mathematics instruction in the lower grades. For his mathematics exam, Alferov faced a committee that consisted of the demoted astronomy teacher, who was out to get revenge on the students who had complained. The teacher gave Alferov a difficult trigonometric equation to solve on the blackboard. Alferov had fortuitously seen the equation and its solution a month earlier in the journal *Mathematics in School*. He hesitated to work out the details on the blackboard quickly, both to recall the problem fully and to give the teacher a fleeting sense that Alferov would fail and he would get his revenge. Then Alferov solved the equation on the board in 15 minutes. The teacher was crushed and lost further interest in the exam, instead asking Alferov what he planned to do with his career. Alferov told him of his desire to become an engineer. The teacher insisted that Alferov was bound for the mathematics-mechanics department of a university, and for a career as a mathematician. Alferov was one of three classmates who got gold medals that year.

The boys socialized with girls but were not serious about them yet. Across the street from the Krapiva house stood a nice one-story house surrounded by a fence in which one of the secretaries of the Communist Party of Belarus, Nikolai Efremovich Avximovich, lived. Avximovich had two daughters. Zoya, the elder, was a student at the nearby 28th Girls' School. Igor Atrakhovich and Zoya met on the way to school and became fast friends. Zoya had a good friend, a pretty blonde named Galya Zhdanovich, who lived nearby. Since Igor and Zhores spent a lot of time together, it was natural for the four of them to get together. Sometimes Zhores and Galya went out by themselves. But Zhores was more interested in his studies than in dating.

In the late 1940s and the early 1950s, when Alferov was in high school and then at the university, anti-Semitism took an official and dangerous turn. Official publications decried "cosmopolitanism" and used other code words to attack perceived Western influences, and Jews were suspected of Zionism. Were Jews citizens of the USSR, or were they naturally drawn to the newly formed state of Israel? In spite of the high percentage of Jews among officials and elsewhere among the intelligentsia, the Party leadership suddenly announced that leading Kremlin doctors, most of whom were Jewish, had been arrested for an alleged plot to poison Stalin and others. Soon other Jews in other specializations and in other cities were also arrested for their alleged involvement in the "Doctors' Plot." A full-fledged terror probably would have consumed tens of thousands of innocent victims had not Stalin died in 1953. The purge ended with doctors pardoned and released, but not before the lives and careers of thousands of people had been ruined.[22]

It was in that environment, Alferov says, that he first noticed Soviet anti-Semitism. He claims never to have suffered its consequences. Though half Jewish, Zhores does not consider himself Jewish; he always says he is an atheist. Yet he says he was disturbed when he saw scrawled in bathrooms, trams, and even one day into the icy window of a tram car the illiterate graffiti of a fellow citizen urging his countrymen to "Fight the Jews, Save Russia!" (Anti-Semitism is growing again today in Russia.)

Alferov has told a series of stories to indicate that his mother's heritage had little place in his life. He points out that his passport listed "Belarusian" under nationality, not "Jewish." (Soviet passports listed one's nationality and served as another check on a person's position in society.) His father, Ivan, ran into official anti-Semitism head on in his dealings with Jewish staff members. The second secretary of the Belarusian Communist Party organization, Vladimir Malin, responsible for industry, called him in for a consultation. Malin and Ivan Alferov first talked briefly about how production was going at the paper trust. Then Malin cryptically asked why so many "Ivanovs" worked at the trust. Alferov eventually sensed that Malin intended to indicate that very few Russians ("Ivanovs") worked in the factory and too many Jews; Alferov employed only one Ivanov, Petr Alexandrovich, the head engineer, Malin continued, pointing out that the head bookkeeper, Ginzburg, the head electrician, Berkovich, and the director of the production administration, Sverzhinskii, all were Jewish. Ivan answered that Ginzburg was the best bookkeeper in the city, had worked at the trust since before the war, and had worked for him in the Urals during the war; that Berkovich was a communist and a leading specialist whom Ivan had brought from the Urals to Minsk, where he continued to be a leading specialist; and that Sverzhinskii had the very highest qualifications, and there was no basis to fire him. Alferov informed Malin that he considered "Leninist principles" in the hiring and firing of employees on the basis of their business acumen and their political reliability. If those principles had been changed or new ones introduced, he would follow them. Malin permitted the matter to drop, and fortunately there were no consequences for Ivan Alferov or for the three Jewish workers.

Alferov Finds Quantum Electronics[23]

Zhores Alferov made his decisions about which university to attend and what career he might pursue in the environment of recovery from war, breakneck rebuilding of heavy industry, growing scientific and philosophical controversies in the sciences, unfolding anti-Semitism, and outspoken support from leaders for scientific research and development. He briefly considered a career in journalism instead of electronics. He mentioned this possibility to the school guidance counselor, whose opinion he strongly respected. She worried that he confused his literary aptitude with a talent for journalism when in fact his experience with journalism was limited to reading newspapers. She informed him, essentially, that it was better to become a good engineer than a mediocre journalist. He chose, of course, electronics. But then came the question of where he should study. One possibility was the newly established Physical Technical Department of Moscow State University (later the Moscow Engineering and Physics Institute, a leading institute of the nuclear age), but Alferov did not consider himself "a wunderkind" capable of studying there, nor did he desire to work in a field that was top secret.

For a variety of reasons, Alferov entered the Moscow Power Engineering Institute (MPEI), studying in the electro-vacuum department under a Professor Kaganov. He chose the MPEI because of his fascination with electronics. Since the first days of the Soviet republic, officials had seen electrification as a panacea. Lenin supported the promulgation of a state electrification plan (known by its Russian acronym, GOELRO) in 1918. Though modest by today's standards, GOELRO was a bold vision for a new state, and it was promulgated 15 years before the US government supported similar programs (such as the Tennessee Valley Authority or the Rural Electrification Administration). Lenin envisioned electric machines transforming backward Russian agriculture, with its backbreaking manual labor and wooden plows, into modern, productive, machine-driven Soviet agriculture, and well-illuminated, well-ventilated factories unlike those in the industrial West. In short, electricity would power socialism.

No less than other high school students interested in electronics, Alferov was enamored of the vision of a modern nation powered by copious amounts of inexpensive electricity. He thought of contributing to electrification by applying himself to the study of long-distance distribution of alternating current based on gas discharge instruments. MPEI, where he determined to study, had been founded in 1930, just a year into Stalin's first five-year plan, to train engineers to design and build thermal and hydroelectric power stations. Its founders proudly claimed that it predated such institutions and programs created during the New Deal in the United States. When Alferov attended MPEI, it was named after Stalin's first deputy, Viacheslav Molotov; later it acquired the name Krzhizhanovskii.

Zhores Alferov and his father went to Moscow and presented Zhores's papers, and Zhores joined the electrovacuum department. (His friend Igor Aleksandrovskii entered the radiotechnical department.) But Zhores did not remain at MPEI long. He detested the dormitory life and the heavy load of purely technical subjects, preferred experiment to the heavy load of theory, and thought himself unprepared for independent life. After discussions with his parents, he returned to Belarus to enter the Energetics Department of the Stalin Belorussian Polytechnical Institute. Of course, neither Molotov nor Stalin had anything to do with electronics, nor was either a scientist. But the Stalinist cult required that their names adorn cities, towns, squares, roads, streets, boulevards, hospitals, schools, and so on.

Because of the familiar surroundings, Zhores found studying in Minsk more pleasant than studying in Moscow, even though the institute's buildings were still under repair and the students often listened to lectures while sitting in roofless halls, wearing overcoats. Zhores liked his fellow students, many of them recently demobilized officers who were quite serious about their studies. For Zhores, physics was fun, and his success helped him overcome the feeling that he had been a deserter for having quit MPEI, especially since Igor Aleksandrovskii was still in Moscow, studying away, and to Igor's family it seemed that Zhores was indeed a quitter. By the summer he had decided to transfer to the department of electronics of the Leningrad Electrotechnical Institute (LETI), about which he had heard much from his high school physics teacher, Iakov Meltserzon.

Zhores's school chum Igor Atrakhovich remained determined to attend the Academy of Arts (later the Repin Institute of Fine Arts) in Leningrad, and had spent a year at a special preparatory school connected with the academy. In 1948 he again applied to the academy. Zhores and Igor traveled to Leningrad to submit their documents, Zhores hoping to enter the sophomore class at the LETI. In his *Memoirs* he wrote: "Thus in summer 1948 I traveled from Minsk to Leningrad where I live to this time. But Belarus will always remain my dear Motherland." His parents did not want him to go to Leningrad, for he was now their only son. His father wondered about his specialization, thinking that boilers and turbines might be crucial to national electrification, but not electronics. "I've never seen an electron," he told Zhores.

Remembering the negative experience of dormitory life in Moscow, Zhores used one of his father's contacts in Leningrad, whom Zhores had met during a family summer vacation, to find quarters. The contact's family had several rooms in a communal apartment on Iakubovich Street, not far from the St. Isaac's Cathedral, where Zhores first lived in Leningrad. He loved the neighborhood and the historical buildings nearby. By 1948 Leningrad's war damage had been largely cleaned up, at least in the center of the city, although there were still many empty lots and skeletons of buildings. Zhores passed these sites on the Number 10 bus from St. Isaac's Cathedral to the LETI, which was located in the historic Petrogradskaia region, near the Botanical Garden and the Neva River.

Two aspects of city life struck Zhores. First, trams and buses, normally stuffed, had room to stand comfortably, if not to sit. Second, the stores were full of high-quality products—sausage, ham, cheeses, yogurt, smoked fish, even caviar. Cans of crab from the Far East were stacked in pyramids in window displays, and cost only three rubles. (Zhores's stipend was 290 rubles per month.) Few people had refrigerators, so wooden boxes affixed outside apartment windows served that purpose. The richness of products may have reflected the inability of most people to afford them, since these goods were available to very few. Indeed, another famine hit the USSR during the postwar years, so Zhores's memory seems faulty on this point.

The Ulianov (Lenin) Leningrad Electrotechnical Institute (today the Saint Petersburg Electrotechnical University) was founded in 1886 as the Engineering College of the Department of Post and Telegraph of Russia of the Ministry of Means of Communication, one of the most dynamic and forward-looking bureaucracies in tsarist Russia. In 1891, as part of a nascent effort to promote the electrification of industry, specialists reorganized the college into the Imperial Institute of Electrical Engineering. Alexander Popov, who developed a radio around the same time as Guglielmo Marconi, was designated the first rector.[24] The institute was a European pioneer in electrical engineering and contributed to the pre- and post-revolution electrification of the country. After the revolution, institute personnel were active in GOELRO, whose first major project, the Volkhovskaia hydroelectric power station, was designed and built under Heinrich Graftiiu.[25] To this day, on trains from St. Petersburg in the direction of Petrozavodsk in Karelia, some passengers look wistfully at Volkhovskaia as a monument of Leninist vision. Leading scientists and engineers connected with such fields as electrical engineering, communications, electric power engineering, electrochemistry, and electrovacuum engineering established strong traditions at the LETI and contributed to plans for the electrification of Russia.

The LETI hosted research on electronics and radio technology through the 1930s. Such leading scientists as Aksel Berg and Mikhail Shatelen passed through its doors, and such professors as Nikolai Borodinskii, Valentin Vologdin, and Boris Kozyrev worked there.[26] In 1933, as at other leading centers, the physicists at the LETI established a new, closed department that focused on nuclear themes. All research and teaching took place under secrecy, access required special permission, and prospective students had their backgrounds checked. Alferov avoided that department, even though he recognized the quality of its instruction and research. After World War II, when Alferov joined the institute, scientists had begun turning to such fields as cybernetics, electrification and automation of industry, telemechanics, and computer science and technology.[27]

Soviet university students received military instruction. At the LETI, in keeping with the fact that Leningrad was a port, there was naval training, which Zhores was required to take as if he were still a first-year student. His Uncle Valya, who lived nearby and was a navy veteran, took Zhores under his wing and instructed him from a textbook on naval matters. One of Zhores's favorite teachers in the Naval Department was Gavril

Aleksandrovich Rumiantsev. Before the revolution, Rumiantsev had served on the *Andrei Pervozvannyi*, an oceanographic research vessel that had toured the North Atlantic and the Barents Sea. Rumiantsev next served in the navy after the revolution, rising to level of admiral. In 1937 he was arrested on trumped-up charges, but in 1940 he was released to serve as a lieutenant again. In spite of their good relationship, Zhores nearly stumbled in his exam on naval technology when Rumiantsev asked a series of questions on topics that only people who had served in the navy could possibly have known the answers. Zhores had two pieces of good fortune. First, the questions required, like a military drill, either a "yes" or a "no" answer. Second, as was usual during the examinations, all the students were in the auditorium and could see how their comrades were doing. Yuri Orlov, an acquaintance of Zhores and an experienced sailor, sat in the audience and nodded or shook his head; Alferov answered accordingly and passed.

Alferov had one last memorable encounter with Rumiantsev in 1957. He had finished his studies at the LETI and was working at the LFTI, where he was also a member of the Komsomol. To raise money for the Komsomol sponsorship of the International Festival of Youth in Moscow in April, the government held a national lottery. Alferov, Aleksandr Lebedev (son of Academician Aleksandr Aleksievich Lebedev, a specialist at the State Optical Institute), and Aleksandr Borshchevskii approached the Komsomol organization at the institute and pleaded for permission to sell the lottery tickets throughout the city. Given a driver and an automobile, they went from market to market, attracting attention by playing records of popular songs through a loudspeaker. They were selling tickets at the Andreevskii Market when up strolled Rumiantsev. Shaking his head, he said: "Alferov. I imagined a brilliant future for you. But here you are . . . 'Buy your ticket!' . . . Phooo." Then he walked off.

Early in his studies at the LETI, Alferov worried that his preparation and training in Minsk would slow him down in comparison with students from Moscow and Leningrad. The centralization of political and economic power in the USSR extended almost logically from Bolshevism to the determination to locate and support the major artistic, educational, and scientific institutions in the major cities, especially Moscow and Leningrad. Simultaneously, personnel in provincial and republican scientific, educational, and other institutions lobbied for increased funding and managed to establish new facilities; for example, in the 1950s a series of republican academies of science were established.

Centralization also extended to national standards for courses, so that curricula differed less from one university to the next than in Europe or

Zhores Alferov in the village of Khil'ki (Kiev region), in 1956, at the monument to soldiers who fell in the Battle of Korsunsk, including his beloved brother Marx.

the United States. And it extended to the quality of institutions of all sorts, those closest to the Party elite in Moscow and Leningrad generally having better instructors, facilities, and other resources. Because of his worry that he might lag in preparation, Alferov studied constantly. During his first quizzes in mathematics, he was surprised to find that he was able to solve all the differential equations, not in the two hours allowed, but in ten minutes. He checked the results for another ten minutes, then turned his paper in. From that point on, he became again a leading authority in the eyes of the teacher, Varta Aleksandrovna Davtian. He got A's in his courses during in the winter session, and he began to gain self-confidence in all subjects. His future technique for research was beginning to be established: he read everything he could find on a subject for background, then worked on problem sets, and eventually began to seek out problems of his own to solve. Later his colleagues would comment on the appearance of his office, in which tireless study gave way to lengthy compilations of calculations and then to the production of detailed hand-written diagrams that covered every open space.

Hero Projects

Early in the autumn of 1949, the entire sophomore class of Zhores Alferov's university visited the construction site of the Krasnoborsk hydropower station in the Oiatsk region of Leningrad province, an area that had been part of Finland before the war. Many descendants of an early Finnish tribe, the Vepsians, inhabited the region near Krasnoborsk adjacent to the Ladoga and Valgjärv Lakes. Alferov was curious about the design of their houses, in which the people had rooms on the second floor while animals occupied the first floor. Officials worried about the "backwardness" of these people, were determined to make them into conscious and productive Soviet citizens, and perhaps even feared their potential aspirations for autonomy; under the policy of "Russification" the number of Vepsians decreased fourfold.[28]

Zhores's sophomore class was part of the nascent student construction brigade movement. In the 1940s and the 1950s, the movement took advantage of sincere interest in building—and rebuilding—the communist state, and paid the students little beyond room and board. One of the main foci of the construction brigades was electrification projects. The LETI students helped design the power station and worked in several brigades in competition with each other to push the project forward. Zhores was in the brigade led by a former front-line soldier, Viktor Ulanov. They dug a diversion channel to draw water away from the dam to the turbine. The project was not significant in size or power, for the Krasnoborsk station was rated at only 100 kilowatts. But it was significant as part of a redoubled effort to bring electrification to rural regions with small hydroelectric and thermal power stations, although rural electrification lagged considerably in the Russian countryside to the end of Soviet power in spite of proclamations to the contrary about how workers and peasants toiled to create socialism. Not content with the Krasnoborsk project, Alferov wanted to see with his own eyes the massive hydropower stations—the "hero projects" (velikie stroiki)—of the late Stalin period.

In 1948 the Communist Party unanimously passed the Stalinist Plan for the Transformation of Nature. According to the plan, which focused on the European USSR, forests, rivers, and fields would be reworked by armies of laborers (many of them gulag prisoners and German prisoners of war) equipped with picks, shovels, wheelbarrows, and eventually modern earth-moving equipment. The goal was to change nature into a machine that operated like clockwork. To many Soviet leaders, nature (in Russian, priroda, a word with a feminine ending) was capricious and arbitrary, like a woman. For too long, droughts and floods had ruined agriculture. Using dams, locks, and hydroelectric power stations on rivers, canals to connect rivers and seas, and irrigation systems in the steppe protected by forest defense belts, they intended to transform nature into a unitary transport, production, and electrical power generation system that would over-fulfill the state's five-year plans. On the Volga River, while the Americans were building such symbols of technological might as the Grand Coulee Dam, Soviet engineers set out to build a dozen power stations, including the massive Kuibyshev station, rated at 1.1 million kilowatts. They engaged in technological competition with the United States, pure and simple, and they often won in terms of speed of construction, cubic yards of earth moved and concrete poured, size of dams, and other measures.[29]

In late September of 1951, at the very end of his vacation, Alferov took advantage of a prize that had been awarded to the five institute students judged to have presented the best scientific papers at a spring student conference. The prize included travel expenses and a coupon for a hotel room for a trip to any Soviet "hero project." Alferov chose to see the Tsimliansk hydroelectric power station on the Don.[30] Its location was important as the transport nexus and power source for the Volga-Don Canal. Why did he select Tsimliansk? He had read in a newspaper about a popular 1948 novel by Vasilii Azhaev, *Daleko ot Moskvy* (*Far from Moscow*), whose main hero, an engineer named Batmanov, was to an extent modeled on the director of the Tsimliansk project. *Far from Moscow*, a typical Socialist Realist novel, glorified the achievements of communist construction, portraying characters who served the state as heroes and those who questioned the infallibility of its leadership as villains. It was a fictionalized account of the efforts of the gulag's railway division to construct a pipeline connecting a new oil refinery at Komsomol'sk to Okha on Sakhalin Island; Azhaev did not mention slave labor. The Far Eastern division of the penal camps, Dalstroi, was in fact an organization geared to using prisoners mercilessly to mine gold and other resources, with disease and death often the results. Azhaev himself had worked in Dalstroi, then had been permitted to have his stories and articles about Dalstroi in the 1930s published and had turned to a career as a writer.[31] In the short stories collected as *Kolyma Tales*, Varlam Shalamov (1907–1982), a prisoner for many years, described life in Dalstroi more accurately.[32] But of course Azhaev, not Shalamov, won a Stalin Prize. And *Far from Moscow* genuinely struck a chord among Soviet audiences— witness Alferov's determination to visit Tsimliansk.

Alferov journeyed to Moscow, where he bought a ticket in a sleeper car on the Moscow-Stalingrad train. While waiting for the night train, he visited the museum of the Battle of Stalingrad. He then took the Stalingrad-Kharkiv train to Morozovskaia and hopped a freight car to Tsimliansk. He arrived very early in the morning at the massive construction project to see huge cranes and excavators seemingly reshaping the earth to the horizon. In the construction village he was able to get a hotel room, which he shared with a captain in the secret police (the NKVD). During the Soviet period, one often shared a hotel room with a complete stranger; this was not out of the ordinary, although such a trip to an unfinished construction site for a university student was unusual, and rooming with an NKVD captain surely encouraged Alferov to watch his words. The next morning, Alferov dropped in to see the director of construction, one Comrade Vanin, yet another individual who had met Alferov's father and knew him as a highly decorated civil war hero. Vanin gave Zhores Alferov a tour of the "hero project."

No doubt Vanin explained how Tsimliansk was typical of "hero projects": it was a hive of earth-moving energy, concrete-pouring enthusiasm, and workers learning their trade on the spot. Tsimliansk consisted of a construction village with a concrete factory and housing that arose belatedly next to the dam (itself finished in 1953). The dam created a huge reservoir that filled the great bend of the Don in Rostov province, some 160 kilometers upstream. The Tsimliansk dam would provide hydroelectric power and irrigation water for vineyards and other fruit enterprises, while the nearby Volga-Don Shipping Canal made possible transport from one river basin to another. In addition to the usual laborers and those employees who distinguished themselves as "shock workers" or "Stakhanovites," the "hero projects" relied on slave labor from the gulag and also some civilian volunteers, about whom Vanin probably was silent. Cubic meters of excavation and concrete poured were precious indicators of progress. Tsimliansk proudly employed an automated concrete factory, suction dredges, and capacious new excavators built at Uralmashzavod, the Sverdlovsk factory renowned for the T-34 tank that turned the tide against the Germans during the war. Other earth-moving machines served as symbols of Soviet modernity even as the project continued to rely heavily on manual labor. Alferov was especially fascinated with an elevator that had been built to help anadromous fish over the dam. But that elevator and the associated "ladders" failed to save fish, as has happened at dams all over the world. The upstream catch of all fish, including the legendary sturgeon, immediately plummeted five- and sixfold. The Don and Volga fisheries may never recover from the cascade of hydropower stations and industrial pollution, but the dams produce electricity aplenty.

On his return to Leningrad, Alferov fell back into his work but was distracted, like many others, by the intensely political atmosphere of high Stalinism, which led to microscopic scrutiny of each person's activities. Elite students at the leading universities felt special pressure to conform to national political and moral norms that included not only allegiance to philosophical principles of dialectical materialism but also correct political behavior on a personal level, two aspects of which were an expectation that one would inform on others and a prevailing prudishness in sexual mores. On one occasion, after repeated prodding from the deputy secretary of the Party bureau of his department (one Zilberg), Alferov presented a lecture on cosmic rays to a woman voter who lived opposite the institute as part of an election campaign. This was not his specialty, and he had influenza at the time, but he met with the elderly woman over his own objections. But the secretary assumed because of his initial reticence that Alferov had refused to meet with her, and reported him to the Komsomol. This led to a full-fledged investigation, then a closed Komsomol meeting at which Alferov was questioned closely. Alferov explained that he had given the lecture to the old woman, and produced a letter from a doctor confirming that he had had the flu. Zilberg impatiently asked why Alferov hadn't informed him of his completed task. "You didn't ask me," Alferov replied. As a result, Alferov was freed of the responsibility of being a Komsomol political organizer, while Zilberg was censured.

At all levels of Soviet society, a kind of public shaming that required self-criticism became widespread. At another Komsomol meeting, the students voted to exclude a young woman from among their midst for her alleged immorality—in this case the accusations were that she attended parties in the dormitory at which the students drank alcohol and that she had spent a night with her future husband before their marriage was officially registered. Alferov claims to have been the only person who spoke up on her behalf. The parties, he observed at the Komsomol meeting, had occurred on May 1, March 8, November 7, New Year's Eve, and the young woman's birthday—days of major celebrations, most of them national holidays. Alferov asked all the Komsomol members who had not drunk alcohol on the aforementioned holidays to raise their hands. Not one did. Then he asked another woman, a member of the Komsomol and an excellent student, "Did you first have intimate relations with your husband before or after registration?" She answered "Before." Zhores proposed, on the basis of answers to his questions and the shows of hands, that it was necessary to expel everyone from the Komsomol organization. (Unfortunately, a resolution to exclude the young woman from the Komsomol passed by

Zhores Alferov's parents resting in the forest along the road from Minsk to Al'bertin, 1954.

one vote. Years later, Zhores ran into the woman on Nevksii Prospect in St. Petersburg, and she thanked him warmly.)

In his second year of university, Alferov gave a talk to the Student Scientific Society on the pre-revolutionary Moscow physicist A. G. Stoletov and his research on photoelectric phenomena. At that time, in an orgy of Great Russian nationalism, the authorities orchestrated a campaign to get school, university, and academy students and scientists to produce works demonstrating the priority of scholars dating back to the tsarist era in a wide range of fields in science and technology; for their part, LETI scientists asserted that Alexander Popov's invention of the radio preceded that of Guglielmo Marconi. After the students' presentations, the head of the laboratory of electrovacuum technology, Nina Nikolaevna Sozina, invited Alferov to join her laboratory, which specialized in research on thin films that produced evaporation in a vacuum. Sozina's laboratory was in a department headed by Boris Pavlovich Kozyrev, who had received a Stalin Prize in 1950 for creating photoelectron-optical amplifiers. Sozina became Alferov's first real mentor, and to this day he feels obliged to her.

By the middle of Alferov's third year in the institute, John Strong's *Fundamental Concepts of Contemporary Physics* and Abram Ioffe's *Theoretical Foundations of Electrovacuum Technology* had become his constant desktop companions. Sozina focused on semiconductors of PbS, PbSe, and PbTe. The pioneer of this work, Boris Timofeevich Kolomiets at the LFTI, went on to study the interaction of light with chalcogenide glassy semiconductors and various photoelectric phenomena. In this environment, Alferov began experiments with Sozina, and also (in the laboratory of gas discharge instruments) under August Augustovich Potsar.

At the end of his third year, Alferov determined that he had made a mistake in his selection of department and perhaps even in his selection of an institute. He now saw quantum mechanics as the key to understanding the new kinds of semiconductor materials that scientists were trying to create. (Precisely at this time, John Bardeen, William Shockley, and Walter Brattain developed the transistor.) He decided to collaborate more closely with Sozina in examining the photoconductivity of different semiconductor films. He delivered his second major scientific address (his first having won him the trip to Tsimliansk); its topic was "Vacuum Methods of Production of Thin Films." Because student presentations were relatively rare, or perhaps because of its originality, the paper earned Alferov another honorary certificate, this one from the Leningrad provincial Komsomol organization and the Ministry of Higher Education. Zhores Ivanovich Alferov had distinguished himself as a diligent, perceptive, and independent scholar at a young age.

Physics and Politics in the Period of High Stalinism

Zhores Alferov came of scientific age in the period of High Stalinism. After the victory over Germany in World War II, many people expected relaxation of the strict controls over daily life, and an end to the constant worry over enemies at home and hostile propaganda against those abroad that had characterized the 1930s. Indeed, in the first months after the war and through the summer, citizens not only gathered in official parades and fairs to celebrate but also gathered spontaneously on the streets to dance, to sing, and to drink toasts to the future with copious amounts of vodka. If it did not entail dropping their guard against the dangers of capitalism, they might consider as allies and perhaps friends the British and especially the Americans, whom they had met at the Elbe River and with whom they had shared sausages, cigarettes, and strong drink. They anticipated increased investment to rebuild housing and agriculture, and also in the consumer sector, so that workers and peasants could work, eat, and sleep well. They understood that the war's great costs—millions killed, the destruction of the nation's breadbasket, Ukraine, and the leveling of hundreds of towns and cities—required not only dedication to the task but also the marshalling of extensive resources to rebuild the socialist fortress. Yet within months of the end of the war Stalin embarked on a fourth five-year plan that in all ways resembled those of the 1930s, with an emphasis on heavy industry and with inadequate investment in housing and agriculture.

While increasing funding for the sciences, including nuclear weapons research, education, and culture, the authorities determined to cleanse society of perceived Western influences. They pursued a policy of political and cultural repression, not relaxation of standards. They established formalistic criteria to ensure that art, literature, theater, and music reflected state policies of ideological vigilance. They determined to eradicate memory of life in the capitalist countries that had been allies during the war. Millions of Soviet citizens had had contact with Europeans and Americans during the war, many as soldiers and others as people who had been dislocated by the war or had been forced to flee and had found themselves on the other side of the front. They saw with their own eyes that, contrary to Soviet propaganda, people lived better under capitalism than they had been led to believe. In fact, the lowly, exploited European worker lived well in comparison with the worker in the Soviet proletarian paradise. Suddenly, many people who (like the soldiers) should have been treated as heroes were arrested and sent off to camps, accused of anti-Soviet propaganda or espionage precisely because of their contact with the West. Perhaps one-third of the Soviet prisoners of war repatriated to their homeland were immediately interned in the gulag. Alexander Solzhenitsyn, a war correspondent who later gained fame as a critic of the Soviet regime and a dissident, fell into the maelstrom of repression because of a careless remark he had made in a letter home about the quality of life in the West—a letter that was read by censors. His labor camp experience and his horror at the repression led him to investigate the camps and to write his monumental three-volume work *The Gulag Archipelago*.

In simultaneous ideological campaigns, the authorities advanced the view that bourgeois culture had little to offer the new Soviet man and woman and was anathema to the Soviet way of life. Andrei Zhdanov, who had been Stalin's right-hand man on ideological affairs since 1934, led the attack. After the publication in *Pravda* in 1946 of a resolution of the Central Committee of the Communist Party that he had orchestrated, ostensibly to criticize two Leningrad literary magazines, *Zvezda* (*The Star*) and *Leningrad*, that had published works allegedly sympathetic toward bourgeois culture, Zhdanov attacked the writers Anna Akhmatova, Boris Pasternak, and Mikhail Zoshchenko (one of Alferov's favorites) for being anti-Soviet and undermining Socialist Realism. In 1947 Zhdanov continued the assault in a speech in which he declared the world had been divided into two hostilely disposed camps, one socialist and the other capitalist. Stalin, Zhdanov, and other leading officials insisted that artists, writers, composers, and scientists all note the danger of admiration of foreign culture. They called for an end to "kowtowing" before the West. Like "enemies of the people" in the 1930s, "cosmopolitans" and bourgeois sympathizers became dangerous appellations in the 1940s.[33] Although Zhdanov died in 1948, the ideological "Zhdanovshchina" (the effort to reestablish the Party's ideological hegemony after World War II) continued for several years. The filmmaker Sergei Eisenstein, the composers Sergei Prokofiev and Dmitrii Shostakovich, and dozens of scientists were attacked brutally and publicly. In this environment, historical writings about science touted the priority of Russian and Soviet contributions in all fields of research, often ignoring the facts or exaggerating. Scholars who failed to recognize the superiority of all things Russian did so at great risk.

Scientists had anticipated a relaxation of the controls on their activities after the war, and a rebuilding of international contacts that Ioffe and others had worked hard to establish in the 1920s. They, too, were sharply disappointed. They had been actively involved in research that had helped defend the Motherland. Ioffe Institute specialists worked, for example, on anti-mine devices for the navy, and a number of them were spirited off to participate in the burgeoning atomic bomb project. Immediately after the war, as thanks and recognition of their future contributions, the government trebled their salaries. The 220th anniversary of the founding of the Academy of Sciences in the autumn of 1945 served as a celebration of Soviet achievements before a crowd of foreign scholars. Scientists thought they, too, might attend conferences abroad. Yet in the late 1940s they also suddenly faced relentless rebuilding, repression, and re-education. Few were permitted to travel abroad, and the selected few often were chosen because of political rather than scientific credentials. Exchanging reprints with colleagues in Europe or the United States might result in charges of collaborating with the West or revealing state secrets. In one case, in 1946, the medical researchers Nina Kliueva and Grigorii Roskin announced they had discovered an agent that shrank tumors in laboratory rodents. Their research gained international attention. It also generated jealousy at home and concerns among bureaucrats that Kliueva and Roskin had given the West valuable information, and they were forced out of the field.[34]

A kind of scientific autarky slowed the development of many disciplines, with the secret police playing an active role in the isolation of Soviet scholars. The NKVD office in each scientific institute accumulated powers to screen mail, sift through journals and reprints, and delay any reasonable request for scientific cooperation even if sent through official channels. Scientists wanting to hold international conferences that Western scholars might attend faced nearly insurmountable challenges. Western scientists hesitated to experience the bureaucratic nightmares of securing visas and living under constant scrutiny to attend meetings in Russia. More and more of them recognized that what they saw in the USSR would be carefully staged, while others refused to visit the USSR as a political protest against totalitarian repression.

In the late 1940s and the early 1950s, when Zhores Alferov read *Pravda* and *Izvestiia*, he read discussions of battles on the ideological front that affected science. A number of fields experienced stultification of research. Most notorious, the self-taught peasant agronomist Lysenko gained control of genetics. In the culmination of a fifteen-year battle, the Lysenkoists labeled genetics a "bourgeois pseudo-science." The internationally renowned plant biologist Nikolai Vavilov had been arrested and had died of malnutrition in a labor camp, while, in a cruel symbol of the arbitrariness of Stalinism, his brother Sergei continued as director of the Physics Institute of the Academy of Sciences in Moscow and became the Academy's president. Alferov read about a national conference of the All-Union Lenin Academy of Agricultural Sciences at which Lysenko had railed on about the dangers of "Weissmanist-Morganist-Darwinian" sciences and had advanced an

outmoded Lamarckian conception of the inheritance of acquired characteristics and the importance of environmental factors in changing both
genotype and phenotype.[35]

Lysenko's ideas, promising rapid changes in variety and hardiness of
crops and livestock, were commensurate with Soviet ideology in several
ways. Just as the authorities would create a new Soviet man and woman
in one generation by destroying capitalism, Lysenkoists would create new
crops by changing the environment. In addition, the fact that Lysenko was
from a simple peasant background made him seem more reliable than the
geneticists, many of whom had been trained in the tsarist era and were from
the middle and upper classes. Many critics of genetics saw Nikolai Vavilov
as a remnant of the past in his class and his political views, although he
and his brother had welcomed Bolshevik support for science. Only nine of
the 56 speakers at the 1948 conference spoke in support of genetics, and
the conference concluded with a series of resolutions (approved by Stalin)
that led to the removal of references to genetics from textbooks, the closing
of departments of evolutionary biology at universities around the country,
and the firing of a large number of professors and researchers.[36]

The lucky geneticists gained jobs at animal husbandry and plant hy-
bridization farms on the periphery of the empire, having to bide their time
doing animal and plant studies while waiting 15 years for Lysenko to be removed. Others found safety in institutes connected with the atomic bomb
project; for example, physicists established departments dedicated to radiation safety and radiation genetics under the protective umbrella of nuclear
research. But many other biologists were arrested and imprisoned. Only
after the ouster of Khrushchev did Academy of Sciences and Party officials
join in pushing Lysenko into retirement at his experimental station. The
fields of chemistry, physiology, geology, and physics also endured extra-
scientific intrusion in the name of vigilance versus the West that slowed
their development and damaged careers and lives.[37]

Physicists participated in angry disputes about the philosophical implications of quantum mechanics and relativity theory. For adherents to
dialectical materialism—and not only for them—the world was real, objective, and knowable. Science, Marxist science included, was the only tool
that enabled diligent apprehension of the laws and regularities of the real
world. Knowledge of this world on all levels, from the subatomic to the
socio-political, contributed to the steady march toward construction of a
communist society. Ideologues believed that to admit idealism and sub-
jectivism in any sphere of Soviet life, including science itself, would subject the working class and the revolution to great risk. Some philosophers

insisted that physicists introduce the notion of class struggle into their activities somehow. In the minds of the opponents of the new physics, relativity theory and quantum mechanics challenged several of the principles of dialectical materialism. Briefly and simply, these opponents adhered to an anachronistic mechanical view of the universe in which electromagnetic phenomena had to be explained in Newtonian terms, a conundrum that Albert Einstein, Max Planck, and others solved through quantum theory and relativity. The opponents also confused physical relativism and its abandonment of absolute space and time as a universal frame of reference with philosophical relativism. Regarding quantum mechanics, they challenged the probabilistic as opposed to deterministic relationships among light, energy, and motion at the subatomic level that suggested the primacy of the individual in perception, the subjective over the objective, even the denial of reality independent of our existence. They were especially disturbed by Niels Bohr's notion of indeterminacy.[38]

Beyond these philosophical issues, political concerns, concerns about status and power, and anti-Semitism contributed to the controversies in physics. Generally, the proponents of mechanistic views were affiliated with universities—in particular, Moscow State University, where the vehemently anti-relativist physicist Arkady Timiriazev once sarcastically denied ever having suggested that Einstein be shot as a solution to disagreements. The university physicists resented the growing authority of scientists connected with the Academy of Sciences. Many of them also disliked Iakov Frenkel, Vitalii Ginzburg, Lev Landau, and other leading theoreticians who were Jewish. Russian nationalism therefore also played a role in their rejection of relativity, as had German nationalism among anti-relativists in Weimar and Nazi Germany.[39]

Though not subject to as much debilitating interference as were biologists under Lysenko, physicists encountered significant obstacles to doing research. Many of them believed that physics research was an apolitical endeavor, and that so long as they focused on their work there would be no problems for them. In fact, most physicists conducted research without outright interference, kept their minds and research equipment to the grindstone, and were held back only by the typical problems of science: insufficient funding, teaching responsibilities, and so on. Yet how they could consider any field of research to be apolitical in the Stalin era is not clear,[40] for in the early 1930s disputes among a relatively small and initially unimportant group of philosophers spilled out into the public and engaged physicists. Debates over quantum mechanics and relativity theory were the least of the physicists' worries. At an Academy meeting in 1936, Abram

Ioffe faced sharp criticism from Party activists for the alleged failure of his institute to do economically important research in metallurgy, electricity, solid state physics, and other fields. Even fellow physicists criticized Ioffe. Lev Landau, who considered himself a Marxist, publicly decried Ioffe's poor leadership at the 1936 meeting. A year later, Landau was arrested under suspicion of anti-Soviet activity and was imprisoned for more than a year, only to be saved by Petr Kapitsa. The brilliant Landau, considered by Niels Bohr to be one of the most promising young theoreticians, had, in fact, Trotskyist leanings, clearly disliked Stalin and Stalinism, and had even distributed leaflets critical of the regime.[41] In 1934, Kapitsa himself had been denied an exit visa to return to his wife and family in Cambridge, England (where he had spent every academic year since 1923), had been put under house arrest, and had lost two years of his career.[42] Matvei Bronstein, a brilliant young theoretician, was one of at least a dozen specialists to be arrested during the Terror and executed by the NKVD.[43] The field of astrophysics suffered a devastating purge.[44]

The experience of the Nobel laureate Vitalii Ginzburg illustrates the arbitrariness of the Soviet system. While officials might attack one scholar for alleged transgression, another might escape unscathed. Although incredibly he retained his Party membership card until 1991, for Ginzburg personally—as for the USSR—communism was an abomination. Ginzburg ultimately rejected Alferov's view that the nation and its scientific community were better off under Communist Party rule than they would have been under another government. Ginzburg recalled the show trials, the murder of the political opposition, the destruction of the Red Army and Party leadership during the Great Terror—in short, the evils of Communism, whether of the Leninist, the Trotskyist, or the Stalinist variant.

Ginzburg was from an apolitical family. But, influenced by the educational campaigns of the 1930s, he enthusiastically joined the Komsomol in 1937 at the age of 21. He became a Party member in 1942, when he earned his doctorate in Kazan, where the Academy had been evacuated during the war. He joined not to advance his career, he writes, but because he believed in Stalin and the Communist Party. He decided not to return to Moscow after the war because he wished to teach in addition to conducting research. He therefore took a job in the radio department of Nizhegorod University (then Gorky University), where met his future wife, Nina Ivanovna Ermakova.

Nina Ivanovna Ermakova's father had died of starvation in a labor camp. A student in the mechanics department of Moscow State University, she lived with her mother on Arbat Street. They were arrested for their alleged involvement in a plot to assassinate a high-ranking official as he drove by their apartment one day. The charges were absurd. NKVD officials ignored the fact that the window from which the attempt was to be made looked onto the courtyard, not the street. Nina was given amnesty after the war but was forbidden to live in Moscow. She ended up in the village of Bor, along the Volga near Nizhnii Novgorod, where she entered a provincial polytechnical institute. Once she had married Ginzburg, and Ginzburg had returned to the Physics Institute in Moscow, he petitioned the NKVD annually to permit Nina to return to Moscow. Sergei Vavilov, director of the Physics Institute and president of the Academy, and Dmitrii Skobeltsyn, who succeeded Vavilov at the Physics Institute upon his death, endorsed each petition. But not until 1953 was Nina permitted to return to Moscow, and in 1956 she was "rehabilitated" with an official letter from the secret police acknowledging that the apartment window in fact looked onto the courtyard.[45]

Vitalii Ginzburg himself faced multiple problems in life and in research. He was a member of the Party who apparently had lost his "class consciousness," he had married a political prisoner, and he was Jewish, and therefore, almost logically, a cosmopolitan. On October 4, 1947, his birthday, *Literaturnaia Gazeta* published an article condemning "kowtowing before the West" and providing a list of some of the individuals who "kowtowed," including A. P. Zhebrak, an opponent of Lysenko, and Ginzburg, who was accused of being silent about the achievements of Soviet physics. Eleven well-known physicists from the Physics Institute wrote a letter of protest to *Literaturnaia Gazeta*, but the editors declined to publish it. Soon thereafter, the notorious VAK (Higher Attestation Committee) refused to give Ginzburg the title of professor, which he needed to teach at Gorky University.[46]

The officials of the VAK were the ideological guards of Soviet higher education. They approved or rejected the award of an advanced degree. Even after a scholar's thesis or dissertation committee approved his defense, VAK officials might turn it down. The reasons for rejection were usually connected with concerns about the scholar's political reliability. Whether a natural scientist or a humanist, perhaps he hadn't taken enough courses in Marxism-Leninism-Stalinism to satisfy the VAK. The VAK's reputation led many individuals to abandon their plans for a PhD. VAK officials maintained that their criteria were objective and clear; a large number of failed defenses suggest otherwise.

Ginzburg's work on the atomic bomb project also played out against the intrigues of the Stalinist system. When initial calculations carried out by a group of scientists under Iakov Zeldovich at the Institute of Chemical

Physics indicated reason for pessimism, Igor Kurchatov (the director of the bomb project) decided to invite another group from Ginzburg's institute to work on the problem. Kurchatov appointed Igor Tamm group leader, although the inclusion of Tamm raised concerns with the KGB since he had been a Menshevik before the Russian Revolution and his younger brother had been shot in the 1930s. Yet Tamm, a chain-smoking bulldog, felt no qualms about inviting Ginzburg to join the group. In spite of Ginzburg's various "defects," and those of his wife, Tamm considered him a superb physicist. Andrei Sakharov also joined the group, and together they developed the Soviet hydrogen bomb.[47]

When Sakharov and Tamm were sent to Arzamas to continue top-secret work on the hydrogen bomb, Ginzburg remained behind as head of a "support group." Although working on top-secret research, he was allowed to visit his wife. Then in 1951 the atmosphere turned even more xenophobic as the Doctors' Plot began to boil over. The authorities restricted Ginzburg's access to his work, and at the beginning of March 1953 it was quite likely that the authorities were going to order the deportation of thousands of Jews to the gulag, including Ginzburg. Stalin died just then, and the plot unraveled. Ginzburg, like other participants in the hydrogen bomb project, received a special award after the successful test in August of 1953. Thinking things had changed with Stalin's death and Beria's execution in December of 1953, Ginzburg was unable to hold his tongue about the evils of Stalinism. The authorities began to follow him and his wife closely again, and friends avoided them. He was denied permission to travel abroad, he lost access to much information now declared secret, and he was barred from a number of conferences, even at home.[48] Still, he rallied Academy physicists to hold a conference in Moscow in 1955 celebrating the fiftieth anniversary of Einstein's special theory of relativity

In the 1960s, KGB authorities (in 1954 the NKVD became the KGB) finally granted Ginzburg the right to travel abroad a few times with his wife, perhaps because of his age. Later, he again lost that right, perhaps because of his closeness to Sakharov, who had become persona non grata in the USSR for his human rights and antiwar stances. Ginzburg never signed any letters against Sakharov when the latter was exiled to Gorky for his "anti-Soviet activity," as most of the leading physicist-academicians did, and even traveled to Gorky to meet with him twice.[49]

And Ginzburg encountered another well-known feature of the Soviet system that interfered with research: censorship. Ginzburg's experience may explain why Zhores Alferov, as an 18- or 20-year-old researcher, had already determined not to join the nuclear establishment. In work on nuclear weapons or other military topics, censorship or secrecy is understandable. Yet officials carried secrecy too far and, in the opinion of Ginzburg and others, did far more harm than good. They interfered with the rational and free exchange of scientific information without hard-and-fast rules. Sometimes they decided that it was a crime to cite foreign authors in an article. Soviet researchers might also face the charge of espionage for publishing results that Western readers might share. The effect was that correspondence with Western colleagues virtually ceased in all fields.

Regarding physics, the officials—and those physicists who rejected relativity theory and quantum mechanics—determined that physics instruction in higher educational institutions was "divorced fully from dialectical materialism," and that "several books of bourgeois physicists have been translated without critical commentary." The dreaded bourgeois physicists included Niels Bohr, Werner Heisenberg, and Erwin Schrödinger—quantum specialists whose work would inform all of quantum physics. Nikolai Semenov, a future Nobel laureate in chemistry, was accused in 1948 of "kowtowing and servility before the foreigners" for publishing a monograph (*Chain Reactions*, 1934) with warm words in memory of Svante Arrhenius and Jacobus Van 't Hoff. In July of 1952, at a Party meeting of the Presidium of the Academy, Abram Ioffe and others were accused of "idealism" for the assertion that matter and energy were somehow related. The attack on physics spilled out into the public sphere in such journals as *Voprosy filosofii*, although one influential quack, Aleksandr Aleksandrovich Maksimov, had to publish his hateful screed on Einstein in the weekly *Red Navy*. This provoked powerful public protest on the part of Soviet physicists, something that was rare for the period.[50]

In 1948, the opponents of the new physics joined with their allies in the Ministry of Higher Education to schedule a conference on ideological problems facing the field. They intended to conduct business on the model of the Lysenkoists, with resolutions from a national meeting leading to strictures on what was considered good physics. No doubt that would have led to the stultification of several fields. At a series of forums leading up to a national gathering, physicists, philosophers, teachers, and Party officials debated the status of physics. The debates indicated the extent of the dangers that physicists faced. Through the intervention of leading scientists connected with the atomic bomb project, which was nearing successful completion, Ministry and Party officials determined to abandon the physics convocation.

In 1990, in the dining hall of the Academy of Sciences in Moscow, Alferov told me a version of this intervention that I later heard from several other leading Academy physicists; it probably is apocryphal: Late in 1948, several physicists connected with the atomic bomb project grew sufficiently concerned about the impending conference to contact Lavrenty Beria, chief of the secret police and government head of the project. As they sat in Igor Kurchatov's office, one of the physicists telephoned Beria to warn him that a decision to find relativity or quantum theory "bourgeois pseudo-science" or some other such ideological denigration would prevent them from completing a bomb. After all, the equivalence of matter and energy lay at the foundation of nuclear weaponry. Beria asked if this was a threat. Assured that it was not a threat but that to construct a nuclear weapon without modern physics would be a physical impossibility, Beria said he would contact them shortly and hung up the telephone. He called back within an hour to inform Kurchatov and the others that there would not be any physics convocation. In 1953, after the death of Stalin, and just before he was arrested and shot, Beria called one of the physicists to let him know that he had discussed the physicists' warning with Stalin, who had replied: "Tell them to build the bomb. Cancel the congress. We can shoot them all later if we need to." Even without the conference, physicists faced great pressure to conform to the dictates of High Stalinism from opponents of the new physics and from Stalinist ideologues who feared Western influences.

Even Abram Ioffe of the Leningrad Physical Technical Institute faced attacks. In print and in private, hostile ideologues criticized his style of management and his views on relativity and quantum mechanics. In the late 1940s and the 1950s, Ioffe confronted incessant questioning. The reasons for this questioning include the tenor of the Zhdanovshchina, an effort to knock Ioffe down a few pegs from his position as unofficial dean of the physics community, the Cold War atmosphere of ideological scrutiny, and growing anti-Semitism. Hence, Ioffe's 1949 book *Osnovnye predstavleniia sovremennoi fiziki* (*Fundamental Concepts of Contemporary Physics*) was a courageous attempt to demonstrate that relativity theory, quantum mechanics, and other fields enabled physicists to understand subatomic processes. Indirectly, he asserted that these theoretical fields promised significant economic benefits in electronics, metallurgy, communications, and other areas, and that the philosophical issues raised were indeed commensurate with Marxist philosophy.

Ioffe's public effort was courageous for three reasons. First, any defense of the new physics was controversial. Second, Ioffe and his institute had faced harsh criticism in the mid 1930s precisely for their alleged failure to contribute adequately to the glorious industrialization effort. Ioffe's book thus was, to some degree, a belated answer to those critics. Third, as has already been noted, anti-Semitism was bound into the controversies, and Ioffe (a Jew) now felt its impact. Iakov Frenkel, the leading theoretician at the LFTI, was also raked over the coals in a series of meetings at the LFTI, and died in 1952 a broken man at the age of 56.[51] Ignominiously, Ioffe was removed as director of the institute he had founded. The new director, Anton Panteleimonivich Komar, had the separate entrance to Ioffe's personal apartment bricked up, and Ioffe never returned to the LFTI. When Alferov became the institute's director, he converted the office of the KGB representative in the institute, the so-called "first department," into the Museum of Abram Ioffe.

Did the bomb save Soviet physics? Under the direction of Igor Kurchatov, Soviet physicists had designed and tested nuclear weapons five or ten years before American intelligence predicted they would. Kurchatov rapidly assembled a team of leading scientists in the atomic bomb project that, by the time of the successful Soviet test in 1949 in Kazakhstan, had become a massive enterprise with dozens of institutes and laboratories, thousands of employees, and few concerns about resources of personnel, machinery, or equipment. Gulag prisoners were crucial to the enterprise. The Americans asserted that only espionage had enabled the USSR to produce a bomb so rapidly. Espionage played a part, to be sure, in convincing Kurchatov and his bosses that they were on the right track. But it was not crucial to Soviet success. The most important factors were the tradition of nuclear physics in research institutes in Leningrad, Moscow, and Kharkiv that dated back to the discovery of the neutron in 1932, the quality of the scientists connected with the project, and Stalin's commitment to provide whatever resources were necessary—personnel, equipment, uranium—as a state priority. And there was always Beria's reminder to succeed or else.

The success of the Soviet physicists in ending the American monopoly on nuclear weaponry was important on three other counts. First, in the international arena it indicated that the United States no longer could dictate how other countries might behave without engaging in realistic diplomacy. Second, the Cold War nuclear arms race, still nascent, would be hard to stop in view of America's unwillingness to listen to any reasonable Soviet offer (and there were a number of offers in the 1950s) and the Soviet Union's xenophobia and secrecy. Third, the physicists' success contributed to the development of another cult in the USSR: the cult of scientists and engineers, who seemed able to build anything—a canal, a massive hydroelectric power station, a bomb—at any time. Physicists in particular acquired great authority, both among Party leaders and among the citizens. Still, solid state physics and electronics remained Zhores Alferov's passion, and he never considered moving into nuclear physics or weapons research. Even when he learned about Hiroshima in 1945 and about the Soviet bomb in 1949, he reacted with curiosity and concern, not with any desire to change fields.

Alferov Settles at the LFTI

Alferov's first years at the LFTI coincided with the fallout from Ioffe's removal, Stalin's death, and the birth of a new USSR under Nikita Khrushchev. By the time Alferov finished his fifth year of university, he had established a comfortable life as a scientist, although his career path remained uncertain. He earned 550 rubles a month as a part-time engineer. His scholarship stipend was increased that year, so he pulled in another 650 rubles. Thus, he had more than enough to live well by Soviet standards. At the end of 1949 he moved to a new apartment not far from Nevskii Prospect. For 150 rubles a month, he had a small room in a communal apartment in which three families and his landlord lived. His senior thesis explored the production of bismuth telluride photoconductive film and the nature of photoconductivity in that film. Work often kept him in the laboratory late into the evening. He understood that in experimental work there were no short cuts, and that the registration of spectra in the infrared region, especially at low temperatures, required great care.

In May of 1952, Kozyrev, the head of the department in which Alferov and Sozina worked, asked Alferov and some other members of the staff to remain behind to inscribe a graph for a paper he planned to present at an upcoming scientific conference. Alferov had asked one of the women students to go to the movies with him that evening, and to call the department if she succeeded in getting tickets. The telephone was in Kozyrev's office. Alferov asked Kozyrev to summon him should the young woman call. But she called several times, and each time Kozyrev told her "No, he's not here." The next day, she told Alferov that she had called repeatedly until 10 p.m. and gotten the same answer. Alferov was deeply hurt. That evening he reproached Kozyrev for having suspected that he might leave before finishing the task at hand, and for misleading him and his friend. Kozyrev did not answer him. After that, not surprisingly, Alferov found their relations formal and chilly.

In 1952, the national matching of graduates with employment was being carried out at the end of October, somewhat earlier than usual. In the USSR's centralized, top-down process, planners and bosses selected workers

Zhores Alferov in 1959, wearing two medals he received in the course of his distinguished career at the Leningrad Physical Technical Institute.

on the basis of the needs of the state, not the other way around. A graduate might be selected to work in a field far from his or her training or interest, and that was the end of the matter; as a result, job dissatisfaction became widespread. Alferov nearly fell into this trap. His senior thesis defense was scheduled for February of 1953, but a special resolution of the government rescheduled it for December. The country urgently needed electrical and other engineers for various defense projects.

The defense went well, and Alferov was awarded a "Red Diploma." He then anxiously consulted with Nina Nikolaevna Sozina. She told him that with some difficulty she had gotten a position added to her department, since she hoped they would continue to do research together. But he told her of his dream to work at the Ioffe Institute. Fortunately for Alferov, Sozina and her husband (also a physicist) had close contacts at the LFTI, called them, and were relieved to learn that another staff position remained open there. Alferov got his dream job. He wrote his mother immediately with the news. Completing the lengthy and invasive job questionnaire that had become standard during the Stalin period required him to phone home for detailed information about all his relatives, their whereabouts, their foreign-language skills, their time abroad, their Party service, and their class. After dropping off the forms at the LFTI, he went home to Minsk to spend a short vacation with his parents before commencing work at the institute.

3   Research and Reforms

Zhores Alferov commenced his scientific career at one of the world's major centers of solid state physics research on the eve of the death of Joseph Stalin. By all accounts, Alferov was far more concerned with science than with anything else, including politics. Remarkably, he remained focused on his research in the period after Stalin's death, when a succession struggle was eventually won by Nikita Khrushchev. Khrushchev then triggered what became known as the Thaw (after a recent novel of the same name by Ilya Ehrenburg). Khrushchev attacked Stalin for his murderous policies and his "cult of personality," although he did not attack the Communist Party itself, nor as a leading official in Moscow and later as Party secretary in Ukraine had he rejected those policies. During the Thaw, Soviet intellectuals and politicians debated the limits of reform in art, culture, science, and even politics.

Scientists took advantage of Khrushchev's more liberal policies to pursue a series of reforms in the administration and organization of science. They gained greater autonomy to determine the direction of research programs. Physicists essentially put an end to philosophical interference in their work, and they and other scientists were permitted greater contact with Western specialists. Although officials continued to press scientists to see to it that their research had immediate applications to ongoing automation, mechanization, electrification, agricultural, and other campaigns, they began to recognize the value of supporting basic research for its own sake. One sign of this was the dismantling of the technical division of the Soviet Academy of Sciences and the transfer of its institutes to branch industrial ministries. This transfer created problems for Alferov and other Academy leaders—they lost several important research institutes to branch industry, and the broad ministerial gulf between basic and applied research grew wider.

When physicists at the Leningrad Physical Technical Institute (LFTI) developed a series of new semiconductor devices, they often felt it necessary to spend precious personal time convincing industry representatives of the utility of their inventions to communications, railways, the military, and other sectors of the economy. While specialists in the United States could and did move back and forth more regularly between industry, university, and military research and development, and at the very least understood how to promote innovation across the three regions of science, their counterparts in the USSR frequently had to present their inventions to industry in person, and often encountered skepticism about the value of their products among personnel who were focused on short-term targets.

Yet there could be no doubt about the prestige of scientists (especially physicists) and their craft in the Khrushchev era. Achievements in space, nuclear power, and a variety of other fields cemented their stature among leaders and ordinary citizens alike. Especially in the Cold War years, scientists commanded vast resources and public honor. Building on such leading institutes as the LFTI and on the expanding network of military institutes connected with space and nuclear weapons, scientists created conditions for world-class research in a variety of fields.

Alferov commenced his research career in this environment of reform and of newfound appreciation for basic research. By all accounts he kept long hours in the laboratory and left it to other physicists at the LFTI to discuss what the Thaw signified for daily life, politics, and science.

In the mid 1960s, when the work of Alferov and his colleagues drew the attention of Soviet officialdom and foreign scientists, Alferov, then in his early thirties, began to think more about family life and also was drawn into administrative responsibilities. His first marriage was spontaneous, short, and unsuccessful, his second a model of family happiness. In the realm of research, Alferov found the field of research—the quantum electronics of semiconductors—that would occupy him for the rest of his career. Working closely with Vladimir Tuchkevich, director of the solid state physics laboratory, he commenced research on transistors and then on heterostructures. In his Nobel autobiography, Alferov wrote: "I recall my first day at the Physico-Technical Institute on January 30, 1953. I was introduced to my new supervisor, V. M. Tuchkevich, head of a subdivision. It was a very important problem to be solved by our not very big team: creation of germanium diodes and triodes (transistors) on p-n junctions." The number on his ID pass, 429, indicated how many scientific employees had passed through the doors of the LFTI in its 35 years of operation.[1]

Leningrad Briefly Weeps, Reforms Begin

Officials had waited three days to announce Stalin's death. They had worried how the public would respond. Stalin's cruel quarter-century of rule had ended, but under his leadership the nation had become an industrial superpower and had survived World War II as the only rival to the United States. The economic battle between two world systems—capitalist and socialist—engaged the world on military, ideological, political, and social levels. An entire generation had been taught that Stalin was the equal of Lenin. But might the Soviet citizen reject communism for a more humane form of rule? Elsewhere, indeed, workers rejected communism. In the summer and the autumn of 1953, workers in East Germany and Poland revolted against the leadership, and in 1956 a bloody revolt against the socialist leadership and the USSR broke out in Hungary. Three of the men sparring to lead the Soviet Union's Communist Party—Nikita Khrushchev, Viacheslav Molotov, and Georgii Malenkov—underestimated the docility of the Soviet citizen. There would be no similar revolt in the USSR. They worried primarily about the danger to them of a fourth pretender to Stalin's mantle: Lavrenty Beria, chief of the secret police. Khrushchev briefly considered joining with Beria against his rivals, then recognized Beria's aspiration to dictatorship. In June of 1953, Khrushchev, Molotov, and Malenkov ordered Beria arrested in the Kremlin, and in December he was executed. Khrushchev then turned his attention to his other rivals. After criticizing Malenkov's seeming endorsement of reforms (including a kind of Soviet consumerism) as an abandonment of Party principles, he accepted these reforms as his own. Over the next three years, Khrushchev pursued a series of domestic and international reforms that were to have direct and indirect effects on basic research.

At home, Khrushchev attempted to orient the economy more toward the benefit of consumers while modernizing industry, to improve agricultural performance and the diet of the citizenry, to apply science and technology to the production process under the banners of "mechanization" and "automation" so as to free workers from manual labor, and to promote greater openness in society through criticism of Stalin's abandonment of so-called Leninist norms. Khrushchev encouraged a crash housing construction program, in which speed often was more important than quality; later, many people referred to the apartments built under this program as "khrushcheby," a combination of Khrushchev's name and the Russian word for "slum." Most people were happy to have their long-promised apartments at last. In the process of advances at home and in industry, citizens and Party officials rediscovered constructivist visions of the approaching communist future in which modern science played a central role. The reforms extended to culture—artists, writers, musicians, and others addressed previously taboo topics, even criticizing Soviet institutions and their legendary bureaucratization.

In February of 1956, in a special session on the last day of the Twentieth Communist Party Congress, Khrushchev read out a litany of details of Stalin's crimes against the nation before stunned Party members. In the so-called Secret Speech, he condemned the excesses of Stalinism, its "cult of personality," and the suffering of millions of citizens. He revealed that Stalin had ordered the murder of Leningrad Party leader and potential rival Sergei Kirov in his office at the Smolnyi (the Communist Party's Leningrad region headquarters and the site of Lenin's first government) and had then used the murder as a pretext to purge the Leningrad Party organization. Though the text of the four-hour speech was not published in the USSR until decades later, many people learned what Khrushchev had revealed. The Secret Speech, and the other reforms, created an informal group, the "children of the Twentieth Party Congress," who hoped to push the USSR back onto the path of Leninist socialism. The "children" included the economist Abel Aganbegian, the sociologist Tatiana Zaslavskaia, and the future Soviet leader Mikhail Gorbachev. Though Zhores Alferov was of the same generation, he preferred the laboratory to political discussions, and was, according to one acquaintance, a "Soviet voice"—that is, someone who thought the reform movement was more dangerous than good.

Khrushchev in no way intended to abandon the centrally planned economy, one-party rule, or several other features of authoritarian government. In practice, the Thaw was more a pastiche of incongruent thrusts and parries, of incomplete reforms and missteps, than a systematic campaign. In agriculture, for example, Khrushchev unthinkingly advanced a "Virgin Lands" campaign to plow under millions of hectares of land in Kazakhstan and western Siberia. The campaign failed miserably within a few years, owing to shortages of labor and capital. In literature, officials found Boris Pasternak's *Doctor Zhivago*—a magisterial rendering of the personal costs of the Russian revolution—to be anti-Soviet, and did not permit the novel to be published in the USSR. Yet in 1962 Khrushchev allowed Alexander Solzhenitsyn to publish *One Day in the Life of Ivan Denisovich*.

Abroad, Khrushchev promoted radically new understandings in international relations through the concept of peaceful coexistence, according to which the Soviet Union and the United States—and their client states—would engage in economic, ideological, and other competitions to determine which system was better, with socialism inevitably victorious over capitalism. Of course, peaceful coexistence did not prevent the Soviet Union (or the United States) from engaging wholeheartedly in an arms race, or from funding national liberation movements or seeking client states.

What did Khrushchev's reforms mean for science and technology, and for a young man or woman entering the scientific profession? Khrushchev believed that in competition between capitalism and socialism the socialist camp would demonstrate its superiority. He and other national leaders touted achievements in science and technology as indicators of the advantages of the Soviet system, including the organization of science and the tie between science and production, and of rapidly approaching communism, which, in the Third Communist Party Program, promulgated in 1961, Khrushchev promised to deliver to the nation in 1980. (This embarrassed such conservative Party officials as Leonid Brezhnev, who, as 1980 approached without any indication that communism was nearing, had to think up another stage between capitalism and communism: "developed socialism." According to a Soviet joke, it turned out yet one more stage existed on that glorious path: alcoholism.)

Three achievements that seemed to indicate to scientists and officials that socialism was overtaking capitalism were the Obninsk nuclear reactor, the tokamak fusion reactor design, and Sputnik. These achievements contributed to the atmosphere of hubris in which Zhores Alferov came of scientific age.

In 1954, at the previously secret Physics Engineering Institute in Obninsk (in the province of Kuluga, two hours south of Moscow), physicists brought on line a 5,000-kilowatt reactor, a forerunner of the Chernobyl reactor, that produced electricity for the civilian grid. Though small by modern standards of 1,000-megawatt reactors, and unfairly rejected as insignificant by the American scientific community because of its size, the Obninsk reactor enabled the USSR to claim that it was the first nation to use the atom for peaceful purposes for the benefit of humankind, while the United States pursued only military ends. Of course, American physicists, like their Soviet counterparts, expended most of their energy and resources on military ends but were also engaged in peaceful applications; the first American civilian reactor (at Shippenport, Pennsylvania, built by Westinghouse Electric Corporation and capable of producing 80,000 kilowatts) was larger than the Obninsk reactor and entirely for civilian ends, but it was not started up until 1958. The USSR was able to reap propaganda benefits from the Obninsk reactor for several years—especially in view of the fact that President Dwight D. Eisenhower, in a 1953 address to the United Nations, had called for "Atoms for Peace."

The second major achievement of Soviet physicists that demonstrated the vitality of their system was in fusion. The development of the tokamak emboldened Igor Kurchatov, head of the USSR's nuclear weapons program, to push for greater contacts with Western scholars. And even as he continued to direct an institute centrally involved in weapons design, Kurchatov urged his government to seek arms-control agreements with the United States, and Kurchatov himself turned to peaceful projects while seeking openness if not full cooperation in scientific matters. For him, science represented the efforts of scientists of all nations. Only by meeting at conferences and workshops or through exchanges where they discussed cutting-edge ideas and became personally acquainted could trust be built between Soviet and American scholars. Kurchatov joined Andrei Sakharov in urging an end to nuclear testing—especially atmospheric testing, which spread radioactive fallout throughout the northern hemisphere and threatened public health.

Kurchatov engaged in a public campaign to ease restrictions on Soviet-American contacts. He insisted upon the declassification of the Soviet tokamak reactor program, to this day the most promising of several interesting alternative approaches to fusion. At the August 1955 Geneva conference on the Peaceful Uses of Nuclear Energy, the first of four such gatherings, Soviet scholars appeared with a flourish on the international scene, largely because of Kurchatov's efforts. Twenty-five hundred scientists, officials, and journalists, from dozens of countries, gathered for the conference. The physicists delivered scores of papers on a wide variety of topics, including the use of isotopes in medicine and industry, the irradiation of food, various reactor designs, and basic research. Eventually the United Nations published sixteen volumes of materials from the conference.[2] Soviet and American physicists happily discovered that they had pursued similar paths on similar projects and that they had encountered similar obstacles in many fields. But the Americans were surprised by the tokamak design. Kurchatov himself recognized the propaganda value of the tokamak, and in April of 1956 he traveled with Khrushchev to the British nuclear facility at Sellafield. Kurchatov and his boss impressed the British with their openness and with Kurchatov's reports of Soviet achievements in a variety of nuclear fields, especially the tokamak design. Alferov recalls reading about the Geneva conferences in the press and excitedly discussing Soviet scientific successes, but at the time he remained focused on solid state physics.

In October of 1957, to top off these nuclear triumphs, the Soviets launched Sputnik. Though small and relatively unsophisticated, it carried transmitters, broadcasting at 20 and 40 MHz, that enabled citizens to track the satellite. Alferov recalls that many of his countrymen gathered on the streets and in fields with binoculars to follow Sputnik. They wrote poems and songs about Sputnik. In the "space race" that ensued, the Soviets shocked the Americans with the first man and the first woman in space, the first "space walks," and the first two- and three-person crews. Soviet engineers, however, had difficulty developing the technology to land a vehicle softly on the moon, and the Americans got there first. But the billions of dollars and rubles that poured into the space race left no doubt about the authority of physicists in both countries.[3]

Scientific Authority and Dissent

As the "children of the Twentieth Party Congress" and other citizens (including scientists) discussed what "reformism" meant for the construction of communism, some of them grew impatient with its slow pace and turned to dissident activity. Alferov, by his own account and those of others, stayed on the sidelines, focusing on his work and perhaps even expressing some discomfort with the extent of the reforms.

Some of the discussion was triggered by novelists. Vladimir Dudintsev's 1956 novel *Not by Bread Alone* depicts the struggles of an inventor against the Soviet bureaucracy. Leonid Leonov's 1953 novel *The Russian Forest* traces a dispute between two foresters, one of whom hopes to abandon wasteful Soviet practices for scientific ones. Ilya Ehrenburg's 1954 novel *The Thaw* criticizes heavy-handed Soviet bureaucratic methods in all spheres. In December of 1954, a few months after the publication of *The Thaw*, Ehrenburg railed against Socialist Realism for its boring, lifeless, less-than-truthful portrayals of good and evil. (Ehrenburg later criticized Boris Pasternak's novel *Doctor Zhivago*, which had won the 1958 Nobel Prize in literature and had drawn rabid criticism in Party and literary circles, but refused to participate in the public campaign against Pasternak.)

Released prisoners soon added their voices to those of the artistic figures who criticized the Soviet regime for its failure to live up to so-called Leninist principles of democratic rule. Many of the early dissenters were returnees from the Stalinist labor camps or member of their families who wished to document their suffering at the hands of the state and to secure rehabilitation, even if posthumously.[4] Marxist philosophers and then a number of scientists joined a growing chorus of dissent that ultimately gained

Andrei Sakharov as its most prominent voice. As the examples of J. Robert Oppenheimer (who was denied security clearance and ostracized by the US government for his opposition to a crash program to build a hydrogen bomb) and Galileo (who was attacked by the Church) suggest, scientists are inclined to question the existing social and political order. And, without a doubt, physicists and mathematicians were among the leading dissidents in the USSR. But as dissent spread throughout Soviet society, Alferov kept his focus on the laboratory.

Although physicists in the Soviet Union and in the United States were confident that government largesse gave them free reign in determining the direction of research, in fact government funding shaped their programs significantly.[5] It pushed them toward weapons research, led to secrecy restrictions, and even shaped fields that seemed far from weaponry, as histories of the US Atomic Energy Commission and the US Office of Naval Research indicate.[6] American scientists battled officials over control of the agenda, subsidization of expensive untested technologies, and increasing concerns about public safety.[7] Soviet physicists may have had some of these concerns, but the inherently closed system precluded any of them from airing the issues in public. We can be certain that the strictures on their behavior, contacts, and publications were more compelling than those on American physicists. For example, as soon as physicists working on the Manhattan Project realized that they would succeed in building an atomic bomb, they began to debate its geopolitical and moral implications. In June of 1945, scientists at the Metallurgical Laboratory of the University of Chicago produced the Franck Report, in which they strongly advised against using nuclear weapons on Japan without first demonstrating the bomb's power in an uninhabited area. Military organization in the Manhattan Project had "compartmentalized" the scientists to prevent military secrets from reaching the enemy, but also to keep specialists from talking about the ethical aspects of nuclear weapons. Many of the scientists who accepted the bombing of Hiroshima were shocked by the bombing of Nagasaki, which indicated that they might be powerless to control future developments. As a result, they lobbied to ensure civilian control over nuclear power and civilian input into nuclear policy. They created the Federation of American Scientists toward the ends of civilian control, then turned to the international arena, believing that only openness and cooperation with the USSR might forestall an arms race.[8] They joined the "One World or None" movement toward international control of fissile material and nuclear know-how. They actively engaged themselves and the public over issues of morality and science in the *Bulletin of the Atomic Scientists* and other publications. Even during the "McCarthy period" (which saw violation of the US constitution in the name of anti-communism, with blacklisting of actors and directors, the requirement of loyalty oaths, and so on), many scientists refused, on the understanding that a scientific citizen must speak truthfully, to silence themselves in regard to the arms race and the dangers of nuclear proliferation. Granted, the USSR had engaged in espionage and had militarily subjugated Eastern Europe. But the ruining of careers, including those of Oppenheimer and other scientists, remains a dark page in American history.[9]

In the USSR, physicists had few opportunities to engage Party leaders, let alone the citizenry, on issues of morality, the bomb, and the arms race. They were prohibited from forming any organizations outside Party channels and from publishing journals or newspapers without prior permission and strict censorship, and were they to oppose official policy they might be declared "anti-Soviet." "Compartmentalization" and top secrecy characterized the military establishment, with strict control on who entered and left entire cities, not only fenced-in institutes and facilities. Occasionally, leading figures spoke up about arms control, fallout, and other issues; for example, in the 1950s Igor Kurchatov and Andrei Sakharov spoke of the dangers of atmospheric testing and the spread of radioactivity. By the early 1960s a number of Soviet scientists were regular participants in Pugwash conferences, and later several Soviet scientists joined International Physicists for the Prevention of Nuclear War.[10] But most Soviet scientists chose a safer route than public activity or dissent: they immersed themselves more fully in their research, often seeking peaceful instead of military applications. Kurchatov threw himself into work on nuclear power generation in fantastical reactors of various sorts. His deputy, Igor Nikolaevich Golovin, became consumed with fusion. And many other scientists in laboratories across the USSR put their noses to the grindstone of basic research. Alferov appears to have been apolitical in the 1950s, rarely paying attention to political questions. Indeed, a number of people commented that the fact he was the son of a loyal, hard-working member of the Party had a strong hold over him, influencing him to keep silent about de-Stalinization and, in the Brezhnev era, to go along with criticism of dissidents and others more often than not.

At the LFTI, most physicists did not engage in overt political activity. They were content with the reforms moving ahead under Khrushchev, and they sensed greater autonomy for their own programs, which they saw as a reflection of greater openness in Soviet culture generally. In the early 1960s, Alferov and others spent time in cafés talking about politics and culture. Alferov read Ehrenburg's novel *The Thaw* some time after its initial appearance. He recalls discussions of the reforms and of anticipated improvements in Soviet life after the Stalin era. He was an active participant in debates with his colleagues over who better understood the human condition: the scientists ("fiziki"), who had produced nuclear reactors, Sputnik, and other great achievements in recent years, or the humanists ("liriki"), without whom morality and other aspects of life could not be properly addressed. Ilya Ehrenburg had published an article in *Komsomolskaia Pravda* that ostensibly was about a failed love affair. A woman had written a letter to the newspaper to complain that her boyfriend, an engineer, was consumed with science and found emotions to be a drag on achievement. A well-known poet, Boris Slutskii, then published a poem, titled "Fiziki-Liriki,"[11] that highlighted the widespread belief that engineers had more to offer society than poets:

For some reason physicists are deeply honored,
While lyricists have been forgotten.
This isn't a matter of dry calculation,
But a matter of universal law.
Indeed we haven't discovered
Something that we really ought to.
Our saccharine iambic poems
Do not provide good wings for us to fly.
And our steeds do not take off on the flight of Pegasus.
That's why physicists are deeply honored,
While lyricists have been forgotten.
It's self-evidently true.
It's useless to argue.
It's not even insulting,
But rather more interesting to observe
Our rhymes disappear like foam.
And their greatness
Calmly gives way to logarithms.[12]

In Western Europe and in the United States, similar debates turned on whether there were "two cultures," as Sir C. P. Snow called them in his now-famous lecture.[13] In the United States, social critics pointed out that landing a man on the moon and bringing him back safely (a goal President John F. Kennedy had set for the nation in 1961) was a simpler matter than solving poverty here on earth. The faith among Soviet scientists and citizens that they would build communism in their lifetimes would remain rather strong until the 1970s, when it would become clear to most citizens that the promised future would be a very, very long time coming. But scientists continued to be revered in Soviet society, and physicists first among them for their contributions to Soviet victories in space and elsewhere.

In 1961, Alferov married. "I thought I loved her and she me, but it turned out not," he recalled. "The marriage was over after three or four months, although it took a year and half to divorce." Alferov lost his apartment in the divorce and had to move temporarily into an uncomfortable basement room at the LFTI. He also had to face a drubbing—orchestrated by his former mother-in-law—in the institute's Party committee. (The brief marriage produced a daughter, Olga, who now works in the Academy of Sciences and whom Alferov sees from time to time.)

In the mid 1960s, Zhores Alferov became acquainted with Tamara Georgievna Darskaia, the daughter of a popular actor associated with the Voronezh Theater of Musical Comedy. Tamara worked at Design Bureau 456, a large, secret space research institute located near Moscow, under Academician Valentin Glushko, who had been arrested during the Great Terror and sent to the Tupolevskaia sharashka, a special gulag camp for scientists.[14] In 1967 Zhores and Tamara married. For the next six months, Zhores commuted to Moscow by airplane to spend weekends with Tamara. Six months later, Tamara moved to Leningrad.[15]

Turmoil within the Walls

In the 1950s, for a variety of reasons, Leningrad was the place for a young physics researcher to be. In quality and quantity of physicists, it ranked with Moscow, the two Cambridges, and the Los Angeles and Chicago areas. When the LFTI personnel who had been moved to Kazan were brought back to Leningrad, at the beginning of 1945, LFTI physicists faced the need to rebuild the entire institute, which had fallen into disrepair from misuse, and to re-establish programs and embark on new directions of research. Work on rectification, contact phenomena, and the photoelectric, thermoelectric, galvanomagnetic, electric, optical, and other properties of semiconductors occupied physicists, as did important projects connected with the burgeoning nuclear enterprise.[16]

The ebbs and flows of political pressure on science at Leningrad State University give a sense of the challenges Zhores Alferov would face in striving to remain quietly apolitical in Vladimir Tuchkevich's laboratory. Under the mathematician Alexander Danilovich Aleksandrov, the university established departments of economics and economic cybernetics. Aleksandrov (a devout communist) defended the scientists from ideological intrusions. As rector, he supported the university's biologists in their struggle against Lysenkoism. He refused Khrushchev's order to re-appoint one of Lysenko's allies, and he protected the teaching of genetics. He permitted students of genetics who had been expelled from other universities to study at Leningrad. He helped reestablish several fields of sociology and economics that had fallen into disrepute among conservative Marxist scholars. Yet political and philosophical turmoil roiled the scientific community through the 1950s. Aleksandrov joined others in worrying about the infection of biologists, chemists, and physicists by the Khrushchev reforms. As rector, he established a new campus for the scientific departments outside of Leningrad in Pavlovsk, largely because the university could not grow any more on its small campus in Vasilevskii Island, but also in part to keep the sciences isolated from the humanities, in which the Thaw touched on political issues and spurred debates about fiziki and liriki and about the direction of future reforms.

At the LFTI, skirmishes over reformism and conservatism threatened personal friendships and allegiances. In October of 1950, the president of the Academy, Sergei Vavilov, called Abram Ioffe to arrange a meeting with him. They had a long, difficult discussion. The details remain unclear, but Vavilov informed Ioffe that he had been fired. Vavilov had tried to defend Ioffe but had been unable to resist "those powerful forces that not only relieved honest people from work, but also several of them sent to the camps." In the Stalin era, not even the Academy's president could defend even such an outstanding scholar as Ioffe—the creator of a physics enterprise that stretched from Leningrad to Ukraine to the Urals. What would happen to Ioffe (would he work at the institute? in his own laboratory?) also remained unanswered. In December, Ioffe sent a letter to the presidium of the Academy "requesting" that he be relieved of the directorship. Ioffe, who had an apartment within the institute with a direct entry from the courtyard, suddenly found that entry bricked up. He was humiliated by having to enter the apartment through the institute's front door, which meant having to show identification. He would come to the institute only one or two more times, and would die in 1960.

In March of 1952, the Academy's presidium agreed to establish a new independent laboratory of semiconductors under Ioffe, authorizing hiring of workers, purchases of equipment and materials, and 300 square meters of space in the building of the former Bekhterev Institute of the Brain. The laboratory opened in May with 36 staff members in three divisions, one devoted to electrical properties of semiconductors, one to thermoelectricity, and one to thermal properties of semiconductors.[17]

In 1954 an Institute of Semiconductors (IPAN) opened under the Academy's presidium, and in 1955 Ioffe became its director with the support of Igor Kurchatov and Anatolii Aleksandrov, two former LFTI employees and now authoritative figures of the nuclear establishment. Ioffe had to be given his own institute, since he had also earned a "Hero of Socialist Labor" award.

Anton Panteleimonovich Komar replaced Ioffe as director. A dry man of Stalinist convictions that probably were reinforced by his rise to the top of his profession from very modest origins, Komar came from a peasant family in Beresna, a village in the Kiev region. At the age of 16 he struck out on his own, working as a laborer, a night watchman, and a laboratory assistant in the physics laboratory of the technical secondary school in Belaia Tserkov. He quickly learned to maintain, operate, and build precision optical instruments and devices. This enabled him to enter the mechanics department of the Kiev Polytechnical Institute. After graduation from the institute, he continued his studies at the Physics Institute in Kiev, then at the LFTI. In 1933, he joined other LFTI physicists in the migration to Sverdlovsk as part of the expansion of Ioffe's physics empire to start up the Urals Physical Technical Institute. His research focused on the physics of metals and included investigations of the structure of metal alloys and of the phase transitions in pure metals and alloys. His work garnered national recognition. In 1946, in Sverdlovsk, Komar and others brought a betatron into operation in a project that grew out of the atomic bomb project and indicated a new focus on high-energy physics. He earned a Stalin Prize for this research, was elected a member of the Ukrainian Academy of Sciences, and then, because of his political reliability and scientific achievements, was appointed deputy director of the Lebedev Institute of Physics.[18]

Taking his responsibilities as director of the LFTI seriously, Komar offended several LFTI traditions. He ran the institute imperiously. He spoke about the lessons of dialectical materialism dogmatically. He earned a reputation for being unforgiving, stern, humorless, and intolerant of small mistakes, and during oral examinations for doctoral and Candidate of Science degrees he often he dwelt on trivialities.

Though there is some debate about the equivalence of Soviet academic degrees with European or American degrees, the Candidate of Science degree often is based on publication of several articles and requires significantly more basic research and writing than a master's degree in the West. Some people doubt that it reaches the same level of formal requirements as the PhD. The Soviet Doctor of Science degree is certainly a more substantial achievement than the PhD, perhaps on the order of being awarded a full professorship. In any event, as was noted earlier, once a scholar had defended either a Candidate of Science or a Doctor of Science degree, the defense committee forwarded its decision to the Moscow-based Higher

Attestation Commission, which vetted each defense one last time to ensure both the scientific and the political reliability of the scholar.

Many scientists suspected Komar of taking special glee in replacing and then belittling Ioffe, just as he enjoyed belittling scholars during their degree defenses. Others believed he was an outsider, even though he had been minted at the institute. They believed he did not give sufficient consideration to solid state physics, the traditional focus of the institute. Although the first nuclear research in the USSR dated to the early 1930s, when it was done at the LFTI under Igor Kurchatov and others, Komar's appointment clearly reflected the growing importance of Cold War physics research and the prestige of nuclear physicists. Under Komar, the nuclear physics program became so large that a branch of the institute was built in Gatchina, south of Leningrad; under the LFTI's next director, Boris Konstantinov, it became the independent Leningrad Institute of Nuclear Physics.

Komar's greatest sin was the willingness with which he conducted a purge at the LFTI. In keeping with the violent, confrontational nature of Stalinist disputation and retribution, he showed Ioffe and others little respect for their imagined transgressions. At least superficially, he conducted business as his predecessor had, focusing on many of the same research questions and keeping the institute's academic council largely intact. Research programs expanded under him. Ioffe was self-important as well, but Komar offended others by too often including himself in nominations for State and Lenin Prizes for research programs in which he had not participated. In personnel policy, Komar seems to have been motivated to do what he thought would please Beria and other bosses.

During the furor over "cosmopolitanism," Komar undertook a campaign against Jews without hesitation. Jews had occupied important positions in leading institutes of mathematics and physics. In absolute numbers, Jews were over-represented among the personnel of the LFTI, while its main institutional rival, the State Optical Institute (Gosudarstvennyi Opticheskii Institut, sometimes amusingly referred to as "goy") was dominated by Russian physicists and was considered "Russian." At this time, Iakov Frenkel—a theoretician of international reputation, with path-breaking publications on the kinetic theory of liquids, quantum theory of metals and excitons, and with a Stalin Prize—faced blistering criticism for his incautious criticisms of dialectical materialism. Dmitri Nasledov (the deputy director of the institute) and Ioffe attacked Frenkel, who was thinking of moving to Moscow for refuge in Semenov's chemistry institute when he died suddenly in 1952.[19] Many people attribute his death to the stress of working under such scrutiny during an orchestrated attack on Jews.

In 1957 Komar was ousted, perhaps because the cleansing of the institute was complete, perhaps because of the Khrushchev reforms and because the Academy's leadership recognized that he could easily be removed after Stalin's death, and perhaps because the institute collective demanded someone with a different research profile. Another problem for Komar was that one Georgii Latyshev had received a Stalin Prize for work on isotopes at the LFTI, and several physicists intimated that Latyshev had falsified his results. This was a stain on the records of Komar and Latyshev, although one could not very easily acknowledge a faulty Stalin Prize. Latyshev was relocated to Kazakhstan as director of a newly established Institute of Nuclear Physics, so his punishment was not crushing. The leadership of the Academy's division of physical-mathematical sciences voted 26 to 19 against extending Komar's tenure as director. He lived out the rest of his scientific life quietly, and few details about his accomplishments after the LFTI are known.

In spite of these intrigues, the LFTI grew rapidly in the early 1950s, largely because of work done for the military. Even with access to the LFTI's archives, it is hard to judge how much funding was connected with the military enterprise, but surely 50 percent and perhaps as much as 75 percent of it was. The Ioffe Institute's programs in nuclear physics grew quite rapidly, necessitating the creation of a campus in Gatchina (mentioned above), which eventually became an independent institute. Boris Pavlovich Konstantinov, director of the LFTI in the early 1960s, had been involved in research on isotopes of light elements used in hydrogen bombs and was crucially involved in securing resources for nuclear programs. At the same time (to judge from memoirs, interviews, and other sources), in spite of the rapid growth of the institute and the evidently large share of secret military R&D, the LFTI remained a "democratic" and collegial place to work, at least on the laboratory level while Komar was in charge, and as a whole once he had departed.

Research on Semiconductors at the LFTI in the 1950s

Vladimir Maksimovich Tuchkevich provided an appropriate atmosphere for research on new semiconductors. Tuchkevich, whom Alferov considers his true teacher and dearest friend[20] and who is regarded among most fiztekhovtsy as an institute builder in the mold of Abram Ioffe for his leadership in expanding the LFTI's programs during his two decades as its director, combined scientific vision, an even-handed managerial style, and a sense of humor. He managed to navigate the challenges of directing 500-some independent-minded physicists during the Brezhnev years of growing conservatism and formalistic efforts to shape research by bureaucratic proclamation and half-hearted reforms. Under the direction of Tuchkevich, the LFTI continued its rapid expansion into plasma physics and maintained its national and international reputation in solid state physics.

The role of a patron in furthering an individual's career may have been greater in Soviet society than in any other industrial society of the twentieth century. Whether in government service, in the Party apparatus, in industry, or in science, one's "chef" facilitated one's entry into programs, one's ascendance through a hierarchy, and one's access to goods, services, and perquisites. Of course, Alferov's scientific talents, vision, and research acumen made it likely that even without such a patron as Tuchkevich he would have moved quickly into advanced research and thence into scientific management. In the case of science, nonetheless, patronage meant entrée into a laboratory, articulation of a stable research program, and the granting of independence so one could explore new avenues. Tuchkevich provided all these things.

Tuchkevich was born on December 29, 1904, in the village of Ianoutsa of Chernygov province, into the family of a teacher. In November of 1919, at the age of 15, he joined the Red Army. While serving in Ukraine he saw the famine that Americans and Europeans helped to end, and from that time on he was an internationalist.[21] In 1924, after demobilization, Tuchkevich entered the physics-mathematics department of Kiev University. Dmitri Nikolaievich Nasledov invited him to join the physical laboratory of the Kiev X-Ray Institute, where Anatolii Aleksandrov, a future president of the Soviet Academy of Sciences, already worked. There Tuchkevich studied the electrical properties of dielectrics under the action of x rays. Like other young physicists from Ukraine, Tuchkevich entered the orbit of the LFTI in 1930 at the All-Union Conference of Physicists in Odessa. In the 1920s, at Ioffe's initiative, physicists banded together to create the Russian Association of Physicists. At roughly biannual meetings that were increasingly well-attended and international (Niels Bohr and Paul Dirac were among the participants), the physicists demonstrated corporate spirit as they discussed new developments in quantum mechanics, solid state physics, and other fields. Their seeming aloofness in regard to Bolshevik intentions to remake society through social and cultural revolution ran directly into the Bolsheviks' determination to hold them accountable to Stalin's "great break" of rapid industrialization and cultural revolution. Subjugated to the Commissariat of Heavy Industry in 1930, the Russian Association of Physicists eventually disappeared.

Ioffe, always seeking to identify young talent to fill his growing empire of institutes in Leningrad, in Kharkiv, and later in Sverdlovsk and Dnepropetrovsk, sought out Nasledov (deputy director of the LFTI under Komar), Aleksandrov, Tuchkevich, and a few other physicists after their presentations at the Odessa conference. Tuchkevich's talk so impressed Ioffe, who happened to be in the audience, that he invited him to join his research team on the spot.

Tuchkevich followed the others to Leningrad in 1936, then worked within the LFTI's walls for five decades. He relished both teaching and research. As a student at Kiev University, he had begun studying the conductivity that arises during the irradiation by x rays of paraffin, sulfur, and other dielectrics. X-ray investigations were central to understanding the crystal structure of the solid state and the insulating, conducting, and dielectric properties of various materials. Tuchkevich focused on copper oxide rectifiers. In all his work, Tuchkevich seemed to be able to see beyond experiment to applications and technology.

Most of the LFTI physicists were evacuated to Kazan when the Germans attacked in June of 1941.[22] They were ordered to fortify right bank of the Volga, to dig foxholes, and to build shelters. They had to trudge seven kilometers from the nearest railway stop. They had no idea whether the Germans would succeed in taking Moscow or whether they would succeed in returning to Kazan. They lived in the hut of a peasant woman, sharing it with farm animals. Eventually they were ordered to turn to war-related research. With Alexandrov, Regel, Kurchatov, and others, Tuchkevich worked on developing ways to demagnetize ships; Tuchkevich had responsibility for the Baltic and northern fleets. The group earned the nation's highest award, the Stalin Prize, in 1942. The LFTI changed markedly after the war as a number of leading scholars moved to Moscow to work on nuclear weapons. Tuchkevich was also included in this work, working on a method for isotope separation on the basis of which he defended his doctoral dissertation in 1955. He then returned to his real interests, solid state physics and semiconductors.[23]

In 1949–50, even while engaged in classified research, Tuchkevich found time to conduct a seminar on solid state physics at the Polytechnical Institute. One of the participants in the seminar, S. P. Nikanorov, recalled that Tuchkevich managed the seminar gracefully and democratically, assigning reports and reviews of literature, commenting judiciously, and providing simple yet profound explanations of phenomena. The discussions ranged from the physics of nuclear explosions to the principles of the operation of crystal triodes with p-n junctions. Engineers create a p-n junction by joining together two pieces of semiconductor. In an n-type junction, phosphorus atoms are added to wafers of silicon to donate extra electrons; here n refers to phosphorus's being negative due to an extra electron. In a diode, p-type (positive) and n-type wafers "are fused together by heating the wafers and the doping substances in a furnace."[24] The students also had affection for Tuchkevich because of his social skills, as well as his skills with a tennis racket and at the wheel of a car.[25] Tuchkevich returned to full-time work at the LFTI in 1949, organizing a sector of semiconductor instruments with some of the youngest physicists on the staff: V. I. Stafeev, Iu. V. Shmartsev, V. B. Chelnokov, A. A. Lebedev, and soon Zhores Alferov. In this group, Alferov eventually solved the problem of more efficient production of germanium diodes. Tuchkevich understood from the start that Soviet industry was not nimble or forward looking enough to commence production of the diodes. He also realized that his laboratory would have to produce a variety of instruments based on the diodes to demonstrate their importance to industry. For example, the Tuchkevich group worked with specialists at the All-Union Scientific Research Institute for Direct Current to produce two models of transformers, one for arc welders at 100 amperes and the other for contact welders at 15 kiloamperes. The manufacturing technology was transferred first to the Lenin All-Union Electrical Engineering Institute and then to the Saransk-based Electrorectifier Factory around 1958. By that time, Alferov had developed a device operating at 500 amperes and 600 volts, a record for the time.[26]

After the first planar transistors appeared in Tuchkevich's laboratory and the first radio receiver with transistors was built, at the beginning of 1953, the physicists turned to serial production of transistors, although in limited numbers. The collective was quite productive. Aleksandr Lebedev produced monocrystals of germanium, Alferov created planar semiconductor amplifiers (several of these, which would now be called transistors, sat on a desk at the LFTI into the 1990s and still worked), Anatolii Uvarov carefully measured their parameters, and N. S. Iakovchuk and Aleksandr Novikov built the first transistor receivers. Tuchkevich asked the latter to bring a receiver to the Smolnyi Institute, the seat of the Leningrad Party Committee, where Lenin had established the Soviet government in 1917. The relatively small tubeless radio produced wonderment, and the physicists were asked to produce three more radios as gifts to the Party leaders Beria, Voroshilov, and Molotov. This was not easy, since it required a large number of transistors. The men worked around the clock, Alferov often sleeping in the laboratory and attentively operating the diffusion oven. The heat that built up in the small room made them quite uncomfortable. One sultry summer night,

Iakovchuk and Alferov were toiling away in their underwear. Suddenly they heard the deputy director of the Institute, Dmitrii Nasledov, shutting and locking his office door. Knowing that he would poke his head into the laboratory, they managed to dress in the nick of time. With a slight smirk, Nasledov asked them to finish their work by the middle of June.[27] As soon as the radios were built, Tuchkevich informed Iakovchuk that the two of them would be traveling to Moscow to present the radios. They were put up at Iakor, a house for visiting scholars, where they waited three weeks as Tuchkevich went around trying to get an audience in the Kremlin. The struggle to succeed Stalin was underway. One evening, Tuchkevich returned to the hotel to announce that tomorrow would be their day. Iakovchuk, unable to sleep, turned on the radio intended as a gift for Beria, whom he had met coincidentally during the spring of 1942 in a discussion over how to improve radio communications among the partisans behind German lines. Upon hearing a report that Beria had been arrested as an "enemy of the people," he telephoned Tuchkevich, who first complained about being awakened, then listened to the radio report, then screamed "Let's get to the train station as quickly as possible!" They headed to Leningrad, where they gave one of the radios to A. I. Berg (Iakovchuk's first teacher at the LETI) and another to V. I. Siforov. Tuchkevich later presented the third radio to the president of the Academy, A. N. Nesmeianov.

At the beginning of the 1950s, the research group studied the physical characteristics of alloyed sharp p-n heterojunctions of germanium and silicon monocrystals, junctions with dopants in those materials, and methods of their production. As a result of that work, the research team established a series of physical laws—for example, that junctions with these impurities carried greater voltage potential than sharp heterojunctions made of the same material. They studied the influence of different admixtures on the characteristics of p-n junctions, and found methods to produce non-rectifying contacts. This required a new diffusion technology. By 1961 the Electrorectifier Factory had finished developing a diffusion technology for serial production of uncontrolled silicon rectifiers. The advantage of the technology was that all processes of diffusion of impurities into a silicon film occurred in the air rather than in a vacuum or a special gas medium, resulting in a simpler and less expensive process. Diffusion made it practicable to add boron or phosphorus to the silicon to create the p-n junction. The researchers developed a way to diffuse aluminum into silicon from its oxide in air. The new method shortened the time of process of diffusion and increased the voltage potential from 1,100 to 1,650 volts.[28] A series of other novel applications of semiconductors followed. In 1955 the

Tuchkevich group built a radio receiver powered by solar batteries, and, as has been mentioned, Alferov developed an instrument that permitted rectification up to 500 amperes and 600 volts.[29]

In the period 1957–1959, researchers in Tuchkevich's laboratory advanced the idea of the construction of powerful rectifying apparatuses for the electrolysis industry, which was still using outdated and unreliable mercury rectifiers. Alferov worked on germanium diodes, which the researchers used in a powerful rectifier (2 kilowatts and 200 amperes) that they tested in a chemical combine factory. This was the beginning of the development of high-current transformers using semiconductors. In the years 1964–1966, Tuchkevich directed the production of a powerful high-frequency transformer for use in an apparatus for production of nuclear fuel—the highly sensitive thyristor TB-320. Perhaps of greatest immediate importance for the national economy, they produced devices for long-distance power transmission. In April of 1968, according to the laboratory journal, the first Soviet transformer-inverter for a high-power direct-current line between the Kashira Electrical Power Station and Moscow went into operation, transmitting electricity at 55 kilovolts and 82 amperes.[30]

On the eve of the breakup of the USSR, its overall capacity of electrical power generation was 350 million kilowatts. About 150 million kilowatts of capacity was transmitted through transformers with the help of powerful semiconductor electronics: diodes, transistors, and thyristors. These devices were produced in twenty factories of the Ministry of Electrical Technology Industry, which employed tens of thousands of workers and which produced nearly 90 million devices in 1988. Physicists at the LFTI trace the roots of this industry to Tuchkevich's laboratory, where in 1954 the first such diodes were developed by Zhores Alferov and colleagues. Tuchkevich understood that the new devices should not simply be handed over to industry. Industry leaders might not fully understand their range of applications, would hesitate to produce new technologies if it meant they might not meet short-term plan targets, might not understand their value, and would have difficulty producing reliable components. Tuchkevich thus insisted that his laboratory group simultaneously pursue development and manufacture transformers to convince industry of their utility. This kind of disjunction between scientific advance and industrial assimilation plagued the Soviet economy for all its years.

The physicists faced dual demands on their time that American and European researchers in universities can probably not even imagine. In the early 1950s, the government, finally understanding the economic potential of semiconductor devices, ordered Tuchkevich's laboratory to produce rectifiers for locomotives. Two researchers developed an oven with a silicon carbide heating element and a series of quartz ampoules that served as the diffusion technology for silicon diodes, but the factory officials had no idea what to do with it. Over the course of a year, the physicists had to run to the factory repeatedly before production lines opened. The burden on them was unimaginable, not only because of the time constraints but also because of the poverty of the laboratory. The physicists had to build much of the equipment with their own hands. They succeeded in manufacturing the ovens and vacuum chambers, but the factory's inability to fathom the utility or the promise of the devices limited their production and reliability. Such disincentives to innovation as industry's reluctance to embrace a new process, the primacy of immediate production targets over potentially longer-term increases in output based on new ideas and technologies, and ministerial barriers were endemic in the Soviet system.[31]

Many leading American physicists moved relatively freely between industry, government, and the university. The three physicists who invented the transistor, John Bardeen, Walter Brattain, and William Shockley, worked for Bell Telephone Laboratories. John Bardeen worked for Gulf Oil, then went to graduate school, worked during World War II in naval research, joined Bell Labs after the war, and eventually settled at the University of Illinois.[32] Alferov had the privileged position in the Soviet R&D system of working contentedly in a laboratory. Initially he was not pulled or pushed into the field to serve as a go-between for research and industry. In the late 1950s, industry and military called on Alferov—as they called on researchers in Academy institutes generally—to introduce his solid state devices into the economy. The challenges he and many other scientists faced reveals, therefore, a significant difference—and weakness—of Soviet science: organizational barriers between research, education, and industry.

Still, LFTI physicists overcame resistance in the diffusion of transformer technology. The first p-n-p-n structures were produced at the LFTI in March of 1962, and by the summer of 1963 the Electrorectifier Factory was manufacturing these thyristors, followed over the next few years by new diodes and simistors, all of which work garnered a Lenin Prize in 1966. At Tuchkevich's initiative, and in collaboration with personnel from the Electrorectifier Factory and the Scientific Research Institute of Direct Current, the physicists determined to build semiconductor transformers for power-line transmission. They developed block rectifiers made of diodes for high-power lines in a rather short time. Grekhov recalls walking the streets of Saransk one night in 1964 with Ilya Abramovich Tepman, the factory's main engineer, and Tuchkevich. They talked sadly about the fact that the devices were not finding their way into the distribution network.[33] But rather than give up, Tuchkevich proposed moving further ahead and using thyristors precisely where electricity was transformed for distribution to an entire region—for example, on the high-power lines from the coal-rich Ekibastuz region to Moscow. This required thyristors of great power. Through extensive correspondence with the State Committee on Science and Technology, in letters co-signed by Institute director Boris Konstantinov, Tuchkevich eventually convinced the State Committee and the presidium of the Academy of Sciences to finance the program. The financing included funding for construction of apartment buildings for researchers and others not involved in the program, one of the first times that had happened. By 1967, the physicists in Tuchkevich's laboratory had developed an experimental thyristor bridge for the Kashir-Moscow power line at 130 kilovolts and 50 amperes, and one for the Volgograd-Donbas power line at 800 kilovolts and 360,000 kilowatts. One of the physicists in the group recalled that it cost them much blood and sweat to organize production lines, yet initially output was miserly. It seems that planners abandoned the project to build the Ekibastuz-Moscow power lines when they learned that the reserves of coal at Ekibastuz were not as great as had been thought. Fortunately, the planners were able to use the powerful blocks on power lines to Finland to export energy at 1 million kilowatts and 400 kV.[34]

Despite the constant pressures simultaneously to conduct research and push industry to understand the value of their work, Alferov and the other physicists have only fond memories of working in Tuchkevich's laboratory. One researcher, Raisa Konopleva, recalled the joy of working for Tuchkevich. Konopleva finished her senior thesis in 1954 in Ioffe's laboratory. Her thesis advisors, A. R. Regel and M. M. Bredov, suggested that Tuchkevich be an official opponent at her dissertation defense. Her work focused on the rectifying properties of germanium irradiated by ions of alkali metals, and in Tuchkevich's laboratory physicists were working on the first germanium diodes. Konopleva had heard Tuchkevich lecture on atomic physics in her junior year at the Polytechnical Institute. She recalled that his lectures were so simple yet thorough that if one had an outline of the lecture one didn't even have to glance at the textbook. Students did not fear Tuchkevich's examinations, since he established a friendly atmosphere as if carrying on a discussion. If a student was unsure, Tuchkevich asked probing questions to direct the student in the proper direction. Konopleva therefore gladly accepted the suggestion to ask Tuchkevich to sit on her committee.[35] Tuchkevich then asked Konopleva to work in his laboratory at the LFTI, where he had a few vacancies. An opportunity to work with Ioffe did not materialize, since he remained in disfavor, his new institute was not yet open, and he could not offer any work. In April of 1954, Konopleva commenced research on high-frequency semiconductors in Tuchkevich's laboratory under A. A. Uvarov. Ninety percent of the laboratory's employees were young scholars who had come from higher educational institutes in Leningrad: the university, the LETI, and primarily the Polytechnical Institute. Konopleva's desk in the lab was next to those of Alferov and V. I. Stafeev. Next door, in Aleksandr Lebedev's laboratory, there were several other young physicists. They considered Tuchkevich, then about 50, also to be young.

Tuchkevich gave his laboratory associates complete academic freedom. One could drop in on him anytime, and problems were solved in seminars or while walking with him as he headed to another appointment. The atmosphere was vital, open, and jovial, even though the work was top secret and all laboratory books, graphs, and notes were stored overnight in portable lock boxes that were under the stern gaze of the KGB. People always celebrated each other's birthdays. But play ended when it came time to deliver research reports or to provide more formal reports to the laboratory director, the institute director, and higher government and Party organs.[36] In this environment, Alferov, with his sense of humor, his love of storytelling, and his absolute seriousness toward science, would thrive.

Tuchkevich as Institute Director

By 1967, when Tuchkevich became its director, the LFTI was a huge research center. In addition to solid state physics, an increasing amount of research on plasma physics was underway, and a huge computer center and laboratory complex had been added. The LFTI expanded its contacts and contracts with industry, and during Tuchkevich's two-decade tenure it doubled in size. As the director of a major center of Soviet science, Tuchkevich had to navigate difficult shoals between science, industry, and government. Indeed, during the Brezhnev era, administrators in the Academy and in the Communist Party, both in Leningrad and in Moscow, became much more bureaucratic in their attitudes toward science. Many bureaucrats believed that any signed piece of paper from any higher body concerning the direction or the pace of research, or related to the glories of "developed socialism," carried more weight than the realities of the laboratory. They insisted that scientists engaged in basic research pay more attention to innovation, pressuring them to take some of the responsibility for the failure of industry to respond more rapidly to their discoveries. Tuchkevich, who had worked hard to get industry to produce new semiconductor devices, criticized this attitude as short-sighted.

Tuchkevich, though apolitical in his avoidance of petty machinations, his disdain for the meddling of Party bureaucrats in science, and his scrupulous refusal to engage dialectical materialism in the laboratory, was a political operator who "worked" the local, regional, and national Party, government, and scientific bureaucracies. Alferov, who was to become the Institute's director after Tuchkevich's retirement, learned many of these skills from him. Since line-item funding was inadequate to the task of building the institute, Tuchkevich made frequent visits to the ministries in Moscow, asking ministry personnel to order the factories in their bailiwicks to assist in producing equipment needed for new programs. On other occasions he secured research contracts from them. This meant that the LFTI received "equipment outside of the plan."

As has been noted, Soviet physicists often faced bottlenecks unknown to their European and American colleagues when trying to acquire advanced equipment, and therefore had to design and build much of their advanced equipment. Alferov envied the ability of scientists in Western laboratories to save time by purchasing equipment from suppliers, while the Soviet instrument-building industry lagged significantly. Soviet physicists produced such specialized equipment as magnets, vacuum chambers, electromagnetic pumps, and laser devices for various nuclear projects, from linear accelerators to reactors.[37] But this equipment served as components of large-scale devices in high-priority areas of space, nuclear engineering, and national security. On his rounds of Moscow bureaucracies, Tuchkevich carried LFTI stationery with him so that at any opportunity he could write a letter of guarantee, a kind of contract with a factory. Occasionally this gave rise to financial difficulties, since these letters, written and signed in Moscow, were not registered in the Institute's planning department and therefore technically the Institute could not distribute the funds. In these cases, Tuchkevich ran a gauntlet of offices in the Academy presidium, from the president to the head of the planning-and-accounting department. In one instance, V. N. Ageev witnessed Tuchkevich sprinting three times during the course of ten minutes from the first floor of the presidium building of the Academy of Sciences to the top floor.[38]

In addition to economic pressures, Tuchkevich faced serious challenges running the LFTI because of the political reaction to the Khrushchev reforms that spread through the country under Leonid Brezhnev, which forced him to focus political energies in areas where he would have preferred not to tread. As part of a nationwide crackdown on political dissent and cultural experimentation, Party officials became much more directly involved in science, including the personnel policy and the so-called social programs of the institutes. As soon as Brezhnev and his colleagues ousted Khrushchev, they determined to reverse the momentum of the Thaw, especially in literature, film, drama, and art. When Soviet citizens engaged in protests and asked their government to live up to its constitutional guarantees of freedoms of speech and the press, or to respect treaty obligations that guaranteed human rights, the government responded by treating the acts as treasonous anti-Soviet behavior punishable by imprisonment. The government identified these people not only as dissidents but as criminals. In the 1950s, the 1960s, and the 1970s, several LFTI physicists transgressed against the unwritten and arbitrary limits established by the regime. On a few occasions, the LFTI was found to be a site of "social programs" that the regime found threatening.

One surprising phenomenon of the Khrushchev era was the appearance of bards whose songs often attacked the government or official policy. Through ballads that had no overt political meaning but still often mocked Soviet values and propaganda, the bards attacked the low physical and spiritual quality of Soviet life.[39] Alferov was attracted to their ballads, perhaps because in a way the bards' music mimicked the themes of his favorite satirists, Mikhail Zoshchenko among them. (The government tried to co-opt several of the bards by allowing them to record on the Melodiia label.)

A number of dissidents worked within the walls of various scientific institutes, some of whom led a life on the edge and a few of whom lost their jobs. In view of the democratic atmosphere of the LFTI and physicists' reputation for independence, it is not surprising that the staff of the LFTI welcomed performances by the bards. Alferov remembers performances by Galich and Vysotsky.[40] The most famous bard was Bulat Okudzhava; his father, a top official of the Georgian Communist Party, had been executed in 1937, and his mother had been imprisoned. Forbidden to live in or visit Moscow or Leningrad until 1956, he performed at the Leningrad Polytechnical Institute, across the street from the LFTI, in 1963, attracting quite an audience. Eventually he performed at the LFTI.

Okudzhava was married to Olga Artsimovich, the niece of Lev Artsimovich, an alumnus of the LFTI and a leading physicist. Olga, a student at the Polytechnical Institute, had lived with her mother in a three-story "barracks" built for employees of the LFTI and their families. Olga loved music and loved to dance to rock and jazz. It is not surprising that she fell in love with Okudzhava. Okudzhava, who had quietly moved to Leningrad to avoid the scandals that surrounded his ballads in Moscow, moved into the barracks with Olga and her mother. LFTI director Konstantinov heard about this; apparently someone had complained to the police, who then complained to him that a musician non grata was on the LFTI's property. Konstantinov, who had suffered at the hands of the NKVD and whose brother had been arrested and shot during the purges, was not about to permit the authorities to force his hand. He called in the chairman of the cultural affairs section of the Institute's trade union branch, Arseny Berezin, asked him what the problem was, and ordered him to organize a concert in the Institute. "No one," said Konstantinov, "will ever remember us with unkind words because we gave shelter to the persecuted poet." On a June evening in 1964, with Olga and her mother sitting in the front row beside Konstantinov, Okudzhava strolled up to the microphone, tuned his guitar, and began to sing in his gravely, insistent voice.[41] After this performance, the political authorities and the KGB tried to orchestrate a scandal. But Konstantinov called someone in the police department to defend the Institute and its physicists, and there the matter stopped. Not so at other institutes.

In 1968, the scientists of Akademgorodok (the Siberian city of science, a symbol of the Khrushchev era, and a kind of oasis of academic freedom far from Party control) had engaged in increasingly overt political activity.[42] Believing that Moscow's rules did not apply to them, they organized social clubs in which they talked about politics, philosophy, and even sex, and they held exhibitions of artists whose work was officially condemned. Several scientists had even engaged in a public signature-gathering campaign to demand human rights.

In Akademgorodok, the turning point came when Aleksandr Galich performed to a standing-room-only crowd in a huge lecture hall at the university. To satisfy those who could not get in, huge loudspeakers were set up on the street outside. Galich performed a variety of ballads critical of Soviet power. He drew prolonged cheers for his "Ode to Pasternak," in which he ridiculed the Union of Writers for publishing, on the occasion of Boris Pasternak's death, only a tiny, four-line notice in the weekly *Literaturnaia Gazeta*.

All these activities contributed to the determination of the Brezhnev regime to crack down on dissidents at home and abroad. In August of 1968, Soviet tanks invaded Czechoslovakia to put an end to that country's experiment in "socialism with a human face." Kremlin leaders forced the arrest of Czech communist liberals. In no uncertain terms, Party authorities informed the leading scientific directors at Akademgorodok that they should stick to science.[43]

In 1965 the Procurator General prosecuted the authors Andrei Sinyavsky and Yuli Daniel for anti-Soviet activity and determined to punish them severely. (Okudzhava eventually acknowledged that Daniel had penned the lyrics for some of his songs.) The government presented an eighteen-page indictment, charged the two men under Article 70 of the Russian criminal code with disseminating "slanderous material besmirching the Soviet state and social system," prosecuted them, and then sentenced them to five to seven years of hard labor. Sinyavsky and Daniel refused to plead guilty and claimed that their writing was literature and therefore protected.[44] The extreme sentences triggered a broader dissident movement that attracted Andrei Sakharov and others.

Two events stand out that reveal Tuchkevich's determination to set limits to prevent outside interference in the management of the LFTI. In 1972, twelve years after Abram Ioffe's death, Party officials and scientists determined that the time had come to reunify the Institute of Semiconductors in Shuvalovo and the LFTI. Toward this end, Party officials insisted that a "cardinal purge" of scientists be carried out. This would involve careful reexamination of the personnel folders at Shuvalovo and microscopic evaluation of the physicists' reliability, followed by mass firings for "heterodoxy." This was not a purge such as those of the 1930s and the 1940s, but it certainly might have signaled the end of someone's scientific career.[45] Tuchkevich, risking his own position, refused to conduct the purge and brought all the Shuvalovo physicists into the LFTI without hesitation.

Tuchkevich also faced increasing pressures to develop an amalgamated research program that reflected the interests of competing Party, economic, scientific, and government organizations, and therefore competing understandings of what science was. The LFTI had to fulfill instructions from the Academy's presidium and, at the same time, orders of the Leningrad provincial Party committee (the Obkom, one step below the Central Committee of the Communist Party). Tuchkevich had little trouble finding a common language with Academy bureaucrats, but conflicts often arose with the Obkom because Party officials and scientists judged the LFTI's responsibilities and successes by different criteria. The Obkom frequently sent investigative commissions to the LFTI. Commission members evaluated plan fulfillment by quite formal criteria that might apply in industry but hardly in an institute conducting scientific research. This made the director's job all the more difficult, because the commissions' conclusions frequently conflicted with the instructions of the Academy. One physicist recalled that Tuchkevich "lost a lot of blood from [the visit of] a commission of people's control of the Obkom."[46]

In the 1970s, Tuchkevich stood up to the Obkom leadership on matters of principle. Although the situation improved somewhat, bad feelings
remained between the Institute and the Obkom. Eventually the Leningrad
Obkom secretary, Grigorii Romanov, ordered a full commission to descend
on the LFTI in 1985 to investigate alleged financial improprieties. Romanov, responsible for military issues in the Politburo, was a conservative,
in the mold of such gerontocrats as Brezhnev, who directed the Leningrad
Party organization for a quarter of a century.[47] A number of LFTI physicists later claimed that the commission led to the improvement of scientific and financial discipline in the Institute. But its members arrived right
on the heels of several other academic and government commissions
that had concluded that research programs and patent activities demonstrated "effectiveness." Nineteen commission members nonetheless spent
several months poking around. This obviously affected the work of the
Institute. The commission members determined that the Institute had
committed a series of mistakes, from accounting to lack of fulfillment
of plans to inadequate innovation in the economy, especially regarding
the "Intensification-90" plan. In the 1970s, Leningrad had been a kind of
laboratory for various experiments to bring the achievements of scientific
research into production. Intensification-90 was yet another effort to focus
on the production end of the research-design-production cycle. For LFTI
physicists, however, the commission's presence and conclusions might
mean only greater accountability to economic planning organizations,
increased bureaucratization, and less science.[48]

Tuchkevich and his staff examined the commission's conclusions carefully and prepared a detailed response. They rejected the negative assessment. They pointed to a record of achievements in science and applications
for the economy and for Leningrad region, and argued that the Institute
had fulfilled plans according to Academy of Science standards. They
admitted some difficulties in innovation, but they put the blame on the
ministerial organizations with which the LFTI had contracts, not on the
LFTI itself. The commission ignored Tuchkevich's response. This provoked
an open protest on the part of Tuchkevich, who said, among other things,
that the work of the commission not only wasn't useful but actually had
a negative effect on both basic and applied research in the Institute. He
firmly asserted that charges of violations of accounting laws were mistaken,
and denied any malfeasance. The Obkom responded by imposing fines on
almost every member of the directorate of the LFTI. Tuchkevich steadfastly
forbade any of the directorate's members to pay the fines, and refused to
sign any documents agreeing with the conclusions of the commission. In
the end, the Obkom abandoned the commission's findings.[49]

Submarines and Semiconductors

While Kurchatov pushed for more contact with Western scientists and pursued peaceful nuclear programs, the Soviet Union continued its military buildup. Beyond nuclear weapons, the atom had other military applications. Like the Americans, the Soviets spent billions on nuclear-powered aircraft, with little success, and billions more on nuclear engines for icebreakers, freighters, and especially submarines (several hundred of them, each with two reactors). Alferov was involved in the latter program directly, though briefly.

Like his father before him, Alferov spent time in the shipbuilding city of Severodvinsk (until 1957, Molotovsk). In the mid 1930s, when he was stationed at paper mills in the Arkhangelsk region, Ivan Karpovich Alferov had worked summers in the shipyards. Now his son would make the journey north.

At the end of a two-lane highway, west of Arkhangelsk on the White Sea, Severodvinsk was one of three major locations for submarine building. Closed to foreign guests until quite recently, and accessible only with permission of the Russian Federal Security Service (Federal'naia Sluzhba Bezopasnosti), Severodvinsk produced several hundred submarines, and the reactors for them, at two major shipyards, Zvezdochka (Little Star) and Sevmash (Northern Machine), and fuel rods were loaded, offloaded, and stored there.

Zhores Alferov participated in the development of germanium solid state circuitry for the first generation of submarines. He went to Severodvinsk, at the request of Anatolii Aleksandrov, to install rectifiers of copper oxide on a germanium substrate. In the event of a reactor shutdown, these rectifiers would enable a nearly instantaneous and 100 percent reliable shift to storage batteries to keep the submarine operational. On one occasion, the deputy head of the Naval Command arrived and accompanied Alferov and others onto a submarine to examine the new work. Zhores pointed out the new rectifiers. Confusing "germanium" with "German," the Admiral asked, with some frustration, "So, we couldn't find any Soviet-made ones?"

Aleksandrov was another alumnus of the LFTI, another of Ioffe's "finds" in Ukraine, and an important figure in the Academy scientific hierarchy with whom Alferov worked directly for fifteen years. Aleksandrov's patronage and friendship enabled Alferov to pursue a number of research and educational initiatives that would have been more difficult otherwise. Aleksandrov remains one of the most interesting figures of the physics community, both for his administrative leadership and for his scientific achievements. Like Kurchatov, he became a fixture in the nuclear enterprise. He was forced into the difficult position of replacing Petr Kapitsa as director of the Institute of Physical Problems from 1945 until 1953 after Kapitsa was ousted because of disagreements with Beria over the running of the atomic bomb project; Stalin's death and Beria's arrest and execution enabled Kapitsa to return, and apparently he bore Aleksandrov no grudge. He became director of the Kurchatov Institute for Atomic Energy on Kurchatov's death in 1960. There he reigned over the development of three generations of nuclear submarines, icebreakers, and armaments. His early focus was the nuclear navy; hence his request that Alferov take a business trip to Severodvinsk. Many people, the present author included, also sense Aleksandrov's support for the ill-fated Chernobyl-type reactor. Others, including Alferov, suggest that he lamented the transfer of responsibility for reactor construction and operation from the Ministry of Middle Machine Building (Minsredmash for short—the ministry in charge of nuclear weapons, among other things) to the Ministry of Electrification. That he also served as president of the Academy of Sciences from 1975 until 1986 leaves no doubt that he was a scion of the Brezhnev-era R&D establishment.

Alferov had first heard of Aleksandrov in 1953 in Tuchkevich's laboratory.[50] Tuchkevich and Aleksandrov had studied together in Kiev and, along with Nasledov and Sharavskii, had been invited to the LFTI at the Odessa conference in 1930. In Severodvinsk in October of 1958, after celebrating successful testing of the germanium circuits in the first generation of nuclear submarines, Tuchkevich and the six physicists with whom he was celebrating dropped in on Aleksandrov, who had been expecting only Tuchkevich and had put aside only one bottle of vodka. During the talking and drinking that followed, Alferov was impressed by Aleksandrov's calm and patient demeanor and by his clear affection for his alma matter, the LFTI. Alferov next saw Aleksandrov in 1968 at a banquet celebrating the fiftieth anniversary of the LFTI. A computer-generated portrait of Abram Ioffe was being passed around for all to sign. Alferov was impressed by Aleksandrov's superb memory of events and people, even some that seemed inconsequential to Alferov.

In October of 1975, when Party officials and scientists celebrated the 250th anniversary of the Academy of Sciences, Alferov had already been elected a corresponding member of the Academy. The celebration was held in the Kremlin and followed the usual ritual. First, one of the Academy's vice presidents, V. A. Kotelnikov, gave a welcoming speech. Normally the Academy's president would fulfill this responsibility, but Mstislav Keldysh had decided to retire because of a serious illness, and a successor hadn't been named. Leonid Brezhnev then gave a speech filled with platitudes about the glories of the Communist Party and the achievements of Soviet science. Presentations by a Moscow factory worker, a woman from a collective farm, and a female student from Moscow University followed. Then Aleksandrov read a congratulatory letter from the Academy to the Central Committee. Alferov and his wife, Tamara, sat close to the dais, in the third row, and saw quite clearly when Brezhnev gave thumbs up to Kotelnikov, then to Aleksandrov, and then said something to Minister Podgornyi. Alferov whispered to Tamara that he sensed Aleksandrov had just been appointed president of the Academy.

Aleksandrov's subsequent election to the post of Academy president—the Academy's members having followed Brezhnev's unmistakable prompt—indicates that by the middle of the Brezhnev era leading Party members and scientists shared many goals and in fact were an overlapping elite. Alferov, having become a corresponding member of the Academy in 1972, was a member of this club of fewer than 237 full members and 439 corresponding members. (In all, the Academy had 40,000 employees, nearly 30,000 of them with higher degrees.) Many members of the Academy also shared an unfortunate incremental approach to the consideration of any reforms that might improve the performance of the economy by granting autonomy to managers and specialists. They considered science to be a branch of the economy, no different from other branches—a branch whose functioning might be improved through proclamations that somehow overcame the infamous bureaucratic barriers between basic research, applied science, and the vast if hulking Soviet economy.[51] Here Alferov distinguished himself on several counts. He was not alone in his integrity, his patriotism, or his devotion to the system. But even as a member of that establishment he sought to reform it, first by invigorating the relationship between higher education and research through new organizational forms that bridged ministerial bailiwicks, and second by encouraging a more natural relationship between research and industry (see chapters 5 and 6). His internationalism and his experience abroad surely shaped these views.

Alferov: Scientific Internationalist

Zhores Ivanovich Alferov came of scientific age in the 1950s and the 1960s, a period of reform and counter-reform, growth of the scientific enterprise, the Cold War, and a belief among physicists in their almost unlimited ability to improve the world. The enthusiasm of the laboratory convinced him that the past and present glories of Soviet science were no accident. With proper organization and financial support, the future promised brilliant successes. Unlike a number of conservative administrators, however, he believed strongly that science was an international enterprise whose future depended on openness and on increased opportunities for cooperation. The Geneva conferences of the 1950s and the 1960s stimulated interest in joint research among Soviet and Western physicists. Several of them traveled to the Niels Bohr Institute for Theoretical Physics in Copenhagen, as if to emphasize that Stalinist pronouncements about the dangerous idealism of Western scholars, prominent among them Bohr himself, were a relic. For example, Spartak Timofeeivich Beliaev, a specialist at the Kurchatov Institute, spent nearly a year with Bohr in 1963–64. Alferov took advantage of the increased openness and growing internationalism of science to present papers and conduct research abroad.

Alferov has always been an internationalist in his view of science. He has been abroad dozens of times, probably two dozen times to the United States alone. His travels began in the 1960s, when it was unusual for a scholar to travel abroad unless escorted by a group of scientists and perhaps a KGB official. Travel to the West, or even to the socialist countries of Eastern Europe, had essentially been stopped in the Stalin years. But nearly immediately after Stalin's death, officials of all Academy research institutes began to highlight foreign contacts in their annual reports to the Presidium of the Academy of Sciences. The number of such visits, many of them to Eastern Europe, grew rapidly through the 1950s. And foreign scientists visited the USSR in greater numbers. The Joint Institute for Nuclear Research in Dubna was a major center of collaborative research involving scientists from the USSR's East European allies, from China, from North Korea, and from Cuba.

Alferov claims that the KGB sought to keep him at home and, in the 1970s, put an unofficial five-year moratorium on his traveling abroad in spite of his international stature and his clear devotion to the USSR. He says the KGB did not appreciate his large number of foreign prizes or enjoy his attendance at international conferences, and that it determined to prevent him from getting awards and to deny him the funding to go abroad. But clearly his successes and his scientific stature in the world were important to the authorities, so that any brief moratorium, if it existed, quickly ended.[52] And, after all, he was a member of the Leningrad provincial Party committee (Obkom) and a delegate to a number of Party congresses.

As Vitalii Ginzburg has written, the obstacles to foreign travel and cooperation established by the KGB and the Party apparatus, daunting even to someone of the stature of Ginzburg or Alferov, indicate the irrationality of the Brezhnev regime's policy toward science,[53] at once trying to be the world's leading scientific power and yet isolating Soviet specialists from their Western counterparts. A scientist's application for a foreign trip—consisting of his biography, a detailed form, and a lengthy discussion of the trip and its value to the individual's career, the institute, and the nation—had to be approved and supported by the directors of his institute. The director of the scientist's institute, the secretary of the institute's Party cell, and the chairman of the trade union all had to sign off on the personal "characteristics" of the candidate. Then another institute commission considered the request, which, if approved, led to an appointment with members of another commission of the district Party committee, whose members were staunch, older Party members who probably wondered what was the value of leaving the great socialist Motherland for the West. If all went well, all the documents, signed and sealed by the commissions, were forwarded to foreign departments of the Academy—in the case of a Leningrad physicist to the Leningrad division, and in parallel to the foreign department of the Leningrad Party organization, and then again, finally, all pieces of paper in a large number of copies to the presidium of the Academy in Moscow, in Alferov's case the Department of General Physics and Astronomy. This tedious, invasive, painful process led many scientists to ignore invitations. "I'm not against flying somewhere to participate in a scientific event," Vladimir Perel of the LFTI told a colleague, "but I'm not going to try to persuade every ass that it is useful and necessary."[54]

Alferov realized the value of foreign contacts early on. He studied German in school because of its importance for his scientific work, and continued its study into the corridors of the LFTI. Then the two volumes of *Transistor Technology* (the published proceedings of the Bell Labs symposia for military, university, and industrial personnel held in September of 1951 and in April of 1952) miraculously appeared in the institute library within weeks of the symposia.[55] Were Soviet agents in America that efficient? Alferov and his laboratory colleagues absconded with the volumes. Alferov determined to learn English so as to be able to read "the Bible," as he and his colleagues called those books. He began to take formal lessons in English. (The teacher, Liudmila Nikolaevna Smirnova, disliked Alferov, he claims, until she discovered he was single. After he married and submitted a composition about his lovely bride, the teacher quickly reverted to disliking him.[56] But another version has it that Smirnova, the most respected English teacher in the entire branch of the Leningrad branch of the Academy, expelled Alferov from the group for not completing homework on time.)

By 1967, when he traveled to London for a meeting, Alferov spoke English fairly well. His English speaking and reading improved greatly during his semester at the University of Illinois in 1970. It helped that two or three evenings a week he joined Nick and Kay Holonyak for dinner at their house. Nick, a physicist, was Zhores's host in Illinois. Kay gave him a dictionary of everyday phrases, and Zhores set out to learn twenty words a day. Without relying on a dictionary, he read two books he could not have obtained in the USSR, Robert Kaiser's *Russia: The People and the Power* (1976) and Hedrick Smith's *The Russians* (1977), and found the descriptions of the USSR accurate.[57]

Alferov first visited the United States in August of 1969. The occasion was an international congress on luminescence at the University of Delaware in Newark. He feels indebted to the Hungarian physicist György Szigeti, who suggested to the conference's organizers that they include Alferov in the program. (Szigeti had visited Alferov's laboratory in 1968.) Alferov encountered logistical problems that were all too typical for Soviet scientists. He faced the unhappy prospect of having to turn down the invitation, since the trip cost 700 rubles and his salary was only 400 rubles a month. But Boris Konstantinov, then the LFTI's director and one of the vice presidents of the Academy, called Alferov to his office and told him that he was prepared to pay for the trip, since he considered Alferov's work on heterostructures quite promising and hoped to establish Soviet priority in that area of research. Then the trip nearly fell apart because the Intourist guide who was to accompany the scientists became ill, and she could be replaced only by someone of the right profile—perhaps an English-speaking member of the Party. This was the opening Alferov needed. He went at Intourist's expense as the English speaker.

Alferov's talk on heterostructure semiconductor lasers that functioned at room temperature provoked quite a stir at the Delaware conference, since it suggested that Soviet physicists were several years ahead of American physicists in that area. He received invitations to visit a number of businesses, scientific centers, and laboratories to discuss this work at the LFTI further. He recalls that he did not feel any of the cool breezes of the Cold War during his visit. Dean Mitchell, a physicist from the Office of Naval Research, had spent ten months at the LFTI in 1968. The ONR supported research in a number of fields with direct military applications—radar, sonar, aircraft, hydromechanics, noise-reduction technologies, deep-sea research, and various applications in electronics in fields closely connected to Alferov's.[58] Mitchell and Alferov had become good friends. Alferov visited Mitchell at the Office of Naval Research in Washington, staying in the

Mayflower Hotel. One day he forgot a notebook, returned to the room, and found a man with tinkering with small tape recorder that had been stashed in the wall. The man blanched, but Alferov told him not to worry, since they both had their respective jobs to do. He grabbed his notebook and left.[59]

In 1970, as has been mentioned, Alferov spent six months at the University of Illinois, where he worked with Nick Holonyak, one of the founders of optical electronics. Holonyak work on silicon thyristors was closely related to Alferov's research. In Illinois, Alferov found himself in a vital, interconnected community of scientists pursuing knowledge of the solid state. Among them was one of the inventors of the transistor, John Bardeen. In 1963, Vladimir Tuchkevich and Boris Zakharchenia (one of Alferov's closest friends) had visited Urbana and met Holonyak and Bardeen. Bardeen had visited the USSR in 1958 and 1961.[60] Zakharchenia had told Holonyak of Alferov's work. Building on the tradition of Iakov Frenkel (who spent time at the University of Minnesota in the 1920s) and Abram Ioffe (who frequented Europe in the early 1920s to reestablish contacts with the West and acquire journals and reagents and then spent a semester at Berkeley in 1928), Zakharchenia had gone to the University of Illinois to work with Holonyak, and had also met Richard Feynman, Carl Anderson, Leo Szilard, Bardeen, and others. Tuchkevich subsequently invited Holonyak to the LFTI; he stayed for a month, including two weeks in Alferov's laboratory, during which the two became good friends. Alferov continued the friendship when he took advantage of the USSR Academy of Sciences–US National Academy of Sciences long-term exchange program to study at Holonyak's laboratory at Urbana.[61]

Alferov characterizes his laboratory of the time (circa 1970) as "state of the art" and recalls that he found "Nick's laboratory to be just like mine." The laboratories had developed in parallel in the course of work on parallel problems, and therefore had developed similar techniques for producing various solid state materials: epitaxy, chemical vapor deposition, and so on. A major difference, however, and one of Alferov's constant complaints, was that the USSR lagged far behind the West in instrument building, especially in the building of equipment that scientists could easily order or purchase in the West. Soviet scientists generally had to do more fabrication than their American counterparts. Alferov loved "to work with my own hands." He was proud of the institute's workshop and construction bureau: "We could have built a locomotive."[62]

Exchange programs that permitted Soviet scientists to travel to American laboratories dated to the early 1960s and reflected the Khrushchev-era hopes of many Soviet and American specialists that relations might improve. The exchanges continued through the end of the Soviet period, by the end of which there were a number of standing commissions on specific regions of scientific cooperation, including environmental sciences and magnetohydrodynamics. When relations between the US and the USSR grew cold, the exchanges grew cold, and when the USSR invaded Afghanistan they essentially ceased for several years. Many observers criticized the program for being advantageous to the USSR and costly to the US, since the US was ahead in most areas.[63] But as the experience of Alferov indicates, we cannot characterize scientific exchanges in black and white.

When Alferov asked the National Academy of Sciences to place him in Urbana, NAS officials hesitated. Holonyak, Alferov claims, had a reputation of not liking to accept foreign scientists. They were shocked that in fact Holonyak agreed to Alferov's assignment without hesitation. "The American side invited me with pleasure," Alferov told me, "because they knew we were ahead [in my area of research]."[64]

Alferov posits that his friendship with Holonyak owed as much to Holonyak's Slavic roots as to shared views of life and science. Holonyak, a native Illinoisan and a son of a former Ukrainian coal miner, was John Bardeen's first student; he went on to get degrees in electrical engineering from the University of Illinois. Bardeen had assembled a number of stellar graduate students around him for work in various areas, first offering seminars on such relevant topics as the transistor (dating to the autumn of 1951), then, a year later, establishing a semiconductor laboratory, to which Nick Holonyak moved from electrical engineering in the spring of 1952. Holonyak remembered his first encounter with the transistor, when Bardeen turned on a small music box that operated immediately, indicating just one of many significant advantages over vacuum tube devices.[65] His classmates could not understand why he gave up vacuum tube research for transistor research.

After getting his PhD degree, Holonyak worked at Bell Labs, for the US Army Signal Corps, and for General Electric before returning to Illinois as a professor of electrical and computer engineering. He contributed to diffused-impurity oxide-masked silicon device technology (transistors, p-n-p-n switches, and thyristors) at Bell Labs in 1956, and invented the shorted emitter, which has found use in the wall light dimmer. (A thyristor, or controlled silicon rectifier, resembles a transistor. It acts as a switch. However, when switched on it can pass current in only one direction.) Holonyak then worked on class III-V heterojunctions. In 1962 he developed $GaAs_{1-x}P_x$ and constructed ruby lasers and LEDs. He has published 500 papers and articles and several books.[66] In addition to his many honorary awards and prizes, he is an honorary member of the LFTI (since 1992) and a foreign member of the Russian Academy of Sciences (since 1999), as well as the first recipient of the Global Energy International Prize in April of 2003, whose prize committee Zhores Alferov chaired.

During his time in Illinois, Alferov joined in a research project with Holonyak and his team on "Luminescence, Lasers, Carrier Interaction Effects, and Instabilities in Compound and Elemental Semiconductors." The researchers studied junctions in heterojunctions, mainly in compound semiconductors and primarily using II-V compounds that they grew by vapor or solution processes. They demonstrated laser operation of InGaP and AlGaAsP in this work.[67]

In the formal report he submitted to the Academy of Sciences upon his return to the Soviet Union, Alferov discussed in rich detail his fruitful six-month visit to the United States. He had spent the majority of his time with Holonyak in his rather small laboratory, with eight graduate students, conducting joint research on recombinational radiation in semiconductors and injector lasers. They had studied luminescence of III-IV compounds and lasers built on their basis, and the tunnel effect and tunnel diodes of gallium arsenide ($GaAs_xP_{1-x}$).[68] In addition to several one- or two-day trips, Alferov spent about a month at the end of his stay in the US visiting other laboratories. He visited the Monsanto Corporation in St. Louis, one of major US manufacturers of semiconductor materials, and saw Monsanto's facilities for the production of epitaxial layers of gallium phosphide in a solid solution (GaP_xAs_{1-x}) and light diodes produced on their basis. He traveled to the General Electric Corporation in Schenectady, where research was underway on light diodes in the green region of spectra on the basis of junctions in GaP and heterojunctions of Al_xGa_{1-x}P-GaP, luminescence, and the technology of matrix screens of red light. At RCA's facility in Princeton, Henry Kressel introduced Alferov to RCA's research program on injection heterolasers in AlAs-GaAs, radiational recombination in nitrides of gallium (Sasha Pankov), the technology of production A3-B5 compounds, and solar cells (Paul Rappoport).[69] At Bell Labs and at IBM's T. J. Watson Research Laboratory in Yorktown Heights, New York, Alferov consulted with "Drs. Bloom, Lorents, Budol and others on light diodes of Al_xGa1_xAs, on the optical properties of different solid compounds of A3-B5 semiconductors, the technology of the production of multi-layer structures with heterojunctions of liquid epitaxy in a system of AlAs-Ga-As and so on." Visits to the Massachusetts Institute of Technology, the California Institute of Technology, the University of California at Berkeley, and the Naval Research

Laboratory in Washington completed the whirlwind trip.[70] Never at a loss for words, Alferov had participated in several conferences and had given eight talks on the work of his laboratory in areas of common interest.

At a symposium on electron-ion and laser technology in Boulder, Colorado, Alferov met Herbert Kroemer, with whom he was to share the 2000 Nobel Prize in physics.[71] Alferov and Kroemer published work on lasers based on double heterostructures at roughly the same time, but their perspectives differed: Kroemer is a theoretician, while Alferov is an experimentalist. Kroemer, a bit older than Alferov, had been studying heterostructures longer. Alferov's friend and NAS contact, reported that Kroemer initially refused to meet with a "Soviet scientist." But when Alferov asked Campbell to pass along a suggestion that he meet with a "Russian colleague" named Alferov, Kroemer gladly accepted, and that evening Kroemer invited Alferov to his home. Alferov brought along the customary half-liter bottle of vodka. They ate well in a family setting, and had a great time talking about "who was first in what." For Alferov it was exciting to hear Kroemer acknowledge that Alferov's paper on the laser had appeared a short time before his. The two men became good colleagues, exchanged letters and reprints, and met at subsequent conferences. In 1996 in Sweden, a special symposium was held at which both of them gave keynote addresses. When they were awarded the Nobel Prize, Kroemer was one of the first people to send Alferov congratulations. But Kroemer always turned down Alferov's invitations to visit Russia, and Alferov guesses that one reason may have been Kroemer's memories of the Cold War and the Soviet occupation of his native Germany.[72]

Soviet Prizes Abound, International Recognition Lags

To encourage competition beyond the stimuli of self-motivation and the desire for priority, in the absence of market stimuli, and because of the country's de facto exclusion from international prizes, the Soviet Union created a series of awards, prizes, and medals. The LFTI won its share of these. In 1976 it earned a "Challenge Red Banner" for its achievements in a national "socialist competition" to raise the effectiveness of production through the application of various devices its physicists had produced.[73] For decades, LFTI physicists regularly participated in these competitions. The Soviet system occasionally punished or ignored those deserving of awards who had transgressed against both written and unwritten laws of the state. It took a special order of the Academy of Sciences in 1990 to elect the physicists A. F. Valter, Iu. A. Krutkov, and George Gamow full members of the Academy posthumously and to recognize their exclusion as baseless.

Alferov and his collaborators at the LFTI won both a Lenin Prize and a State Prize. The prize process began with a nomination from a research collective, often the academic council of the institute in which the research took place. The nomination to the Committee on Lenin Prizes or State Prizes in the Area of Science and Technology of the Council of Ministers of the USSR consisted of a cover letter from the institute and a series of supporting documents, including a brief summary, an in-depth discussion of the discovery, résumés of the participants including their publications, and such other materials as written testimonies of specialists in the field, evaluations of the economic impact of the discovery, and a lengthy bibliography. An application might be submitted several times and never win an award.

In the case of the materials presented for the 1965 Lenin Prize on behalf of Vladimir Tuchkevich and four collaborators for work on new semiconductors, the application began with a three-page letter from the LFTI's director, Boris Konstantinov.[74] In the letter, Konstantinov pointed out that research on physical processes in p-n junctions and the development of technological methods to make them from semiconductor materials carried out at the LFTI since the beginning of the 1950s had led to new, powerful silicon diodes, triodes, and other devices. He noted the physicists' efforts to work with the Electrorectifier Factory to develop new diffusion technology that would make it practicable to produce devices serially, including new uncontrolled silicon diodes, rectifiers rated at up to 350 amperes and up to 1,000 volts, and controlled diodes rated at up to 200 amperes and up to 1,500 volts. These devices, Konstantinov indicated, had significant industrial and military applications in non-ferrous metallurgy, in the chemical industry, in railroads and other transport industries, and in energy production that would lead to "radical changes in electrotechnology and several regions of energetics." Economists had already predicted significant savings, with a forecast of up to 500 million rubles in the next five years. Nominated for the prize along with Tuchkevich were Valentin Chelnokov, Igor Grekhov, V. B. Shumana, and I. A. Linichuka.[75]

Tuchkevich and his colleagues provided a lengthy explanation documenting their research program. Semiconductor devices, they asserted, would be crucial to solving a series of problems connected with the generation, distribution, and use of electrical power, since the economy increasingly relied on direct current for railroads, for municipal transportation, for production of titanium, manganese, copper, and aluminum, and other metals, and for production of chlorine and hydrogen. The researchers proudly concluded that "the simplicity of the technology, its effectiveness, the conditions of mass production and the achieved parameters of the instruments have achieved a high appraisal by specialists in a series of socialist countries."[76]

What were the clear advantages of the new devices? Mercury rectifiers were inefficient, short-lived, and sensitive to vibrations and temperature; mechanical ones were unreliable and heavy and required complex controls and highly qualified personnel. Motor generators were huge, complex machines that required a lot of copper and steel and were relatively short-lived.[77] The new semiconductor devices were "extremely reliable even in heavy use," could "withstand shock and vibration," were "hermetic [and] corrosion resistant," and could "handle rapid changes in temperature." They were light and small. The inventors claimed that their operating specifications met or surpassed those of devices produced abroad by GE, AEG, Braun Boveri, Westinghouse, and other major corporations.[78] In addition to several of his earliest publications, Alferov's contributions to the Tuchkevich team in this area of research included two papers filed for Soviet patents: "Powerful Germanium (Silicon) Rectifier with High Shock- and Vibration-Stability" (registered April 16, 1960) and "A Covering of Tungsten with Tin, Indium, Silver, Nickel, Aluminum and other Alloys on their Basis" (registered September 14, 1960). Alferov was also a co-author of a paper titled "Powerful Impulse Germanium Rectifier."[79]

The Tuchkevich group did not receive a prize in 1965, but in 1972 the State Committee awarded a Lenin Prize to a group of LFTI researchers under the direction of Alferov, including Viacheslav Mikhailovich Andreev, Dmitrii Zalmanovich Garbuzov, Vladimir Ilich Korol'kov, Dmitrii Nikolaevich Tretiakov, and Vasilii Ivanovich Shveikin, for "fundamental research on heterostructures in semiconductors and the creation of new instruments on their foundation."[80] This was the very work for which Alferov was later to receive his Nobel Prize. (Over the period 1967–1987, LFTI physicists earned seven Lenin Prizes, 52 State Prizes, and 21 Komsomol Prizes.)

In spite of the importance of Alferov's work for semiconductor physics and its future applications in solar, communications, and other technologies, recognition of his contribution lagged. Fear of releasing state secrets may have been a factor. In any event, the Soviet press finally reported with joy that, at a meeting in Budapest in 1970 on semiconductor physics, Robert Anderson, a professor of materials science at the University of Maryland, expressed the opinion that, although heterostructures had a number of positive characteristics, the creation of instruments on their basis was a far-off dream. The same evening, Zhores Alferov delivered an enthusiastic lecture indicating that the dream would soon be a reality and reporting work on heterojunction lasers and other new instruments done at the LFTI. By 1972, Soviet journalists had begun to play up the promise of these instruments. One wrote: "We have become accustomed to the entry of semiconductors into our life." From transistors, the writer continued, physicists had moved on to solar batteries, rectifiers, barrier layer cells, and thyristors of light weight, small size, and high efficiency based on the achievements of the laboratories of Alferov, Kroemer, and others. The new materials used in the heterojunctions also were mentioned. If in 1960 the goal was to create ideal heterostructures—something Alferov and his group had achieved—the long-term promise of heterostructures depended on practical matters, such as overcoming the corrosiveness of aluminum arsenide. The Alferov group succeeded in producing heterostructures that were stable at room temperature and that served as the basis of new instruments—high-voltage diodes, light sources with high output, photoelements, and so on.[81]

Having made his fundamental contribution to physics, Alferov continued his research on heterojunctions with efforts to improve their parameters, to find other materials for their production, and to promote their industrial manufacture for a variety of applications. He also began to move into the administration of science, first as a corresponding member of the Academy, then as a full member and later the director of the LFTI, as one of eleven vice presidents of the Academy, and as chairman of the Leningrad Scientific Center. In these administrative posts, Alferov learned from the example of Tuchkevich.

The Academy of Sciences, the leading scientific institution in the USSR, was, increasingly, run by aging academicians. Beginning in the mid 1920s, scientists had succeeded in gaining broad public and official support for the creation of new institutes and laboratories. They had maintained a modicum of autonomy in administration, though less so in the social sciences (which were more ideologically suspect) than in the natural sciences. Academy scientists, especially full members ("academicians") and corresponding members, had acquired great prestige in many fields of science. Almost without exception they had published widely. Many of them had secured international reputations, though fewer of them had, like Zhores Alferov, managed to travel abroad for a conference, let alone spend a month or a semester in a foreign laboratory. Yet, perhaps like all bureaucracies, and certainly like the legendary Soviet bureaucracies, the Academy had become increasingly conservative. Its members were well fed by the Communist Party. They saw no reason to disrupt the status quo.

They wished to maintain leadership in all fields of fundamental research, and to have a strong say in the development of other areas of activity, including cosmic research and computer science and technology. The number of full members of the Academy had grown from 45 in 1930 to 300 in 1979, reflecting the expansion of the scientific enterprise generally and the emergence of new fields of research. Some of the Academy's members had become stodgy administrators who intended to maintain preeminence in their fields through control over institutions, personnel appointments, and purse strings. We should therefore not be surprised that Zhores Alferov was not elected to full membership in the Academy until 1979.[82]

By becoming an Academician, Alferov also became a celebrity. He loved giving interviews on television and in the newspapers. (He still does.) In every forum, whenever he spoke, his sincere loyalty to the LFTI, to Leningrad, and to his nation came through. In an extensive interview published in 1979 in the Leningrad evening paper *Smena*, Alferov noted that an entire industry now produced instruments based on heterojunctions: lasers, light diodes, solar photoelements, and so on. "You can say with complete certainty that in the last years our idea has found full realization." The economic significance of heterojunctions would grow as fiber optics, LEDs, and photoelements entered everyday life. Alferov attributed these achievements to the atmosphere at his institute, which was based on the "organic combination of fundamental and applied problems in research, democracy in relations [among scientists] independent of rank, and most important, devotion to one's favorite cause—to physics, the institute, Leningrad, the country." Another constant theme was the importance of youth in science, since the young "have always been the engine of science."[83]

In his public life Alferov strove to provide promising young students with better, more efficient opportunities to get early training and then to move into leading research institutes. Two important characteristics of American science and technology are the healthy relationship between university training and research and the ease with which people in universities and industry can change their venue of employment. In the USSR, as Alferov and many others argued for decades, the bureaucratic and ministerial barriers between education, basic research, and applications slowed developments at nearly every step. First, a kind of gap existed between the training of students in universities, polytechnical institutes, and other higher educational institutions (often referred as Vuzy, a Russian acronym) and basic research in the sciences, which was done primarily in institutes of the Academy of Sciences. Fortunately for the physicists at the LFTI, the Leningrad Polytechnic Institute was directly across the street, and Tuchkevich,

Konstantinov, Alferov, and others taught there in order to attract the best students to full-time research. Second, while Soviet academic institutes increasingly earned income from contracts with branch institutes and their laboratories, where applied research in the sciences generally was done, neither the results of research nor the researchers themselves could move easily between academic and branch institutes. Soviet researchers were far less mobile in their careers than their Western counterparts; indeed, most of them retired from the same institutions in which they had first found employment. And the Soviet Union's artificial methods of promoting innovation through plans simply did not have the dynamism of the West's market-driven innovation.

To this day, Alferov remains angry about Khrushchev's decision to declare battle on the legal system of paying individuals two salaries, one at the main place of work and another at a second job. The battle against sovmestitel'stvo (literally "working at two places") meant that a scientist who was paid at a research institute, say the LFTI, could no longer be paid to teach at a university or a polytechnical institute. That was detrimental to the training of new researchers and to creating some dynamism of the sort that existed in the US scientific enterprise. The LFTI's strong ties to the physical mechanical department at the Polytechnical Institute were nearly broken. Fortunately, in a so-called base department at the Leningrad Electrotechnical Institute, the scientists maintained strong connections that enabled them to identify promising students and then work with them in laboratories, in seminars, and on special projects. They identified promising high school seniors through periodic "olympiads," and they established an early-spring "school" in the LETI's department of electronics on the theme "Physics and Electronics." (It was occasionally conducted on skis in a park.[84]) Semen Grigorievich Konnikov, a leading researcher and educator at the LFTI who arrived from Moscow in the mid 1960s and who first worked with Alferov in 1968, credits Alferov with building scientific camaraderie at the institute. Beginning in mid 1960s, Alferov organized weekend seminars, held at various parks outside Leningrad, in which physicists, young and old, discussed cutting-edge ideas and also relaxed and played sports. "Women and children came along, too," Konnikov told me. "This 'hetero-empire' brought us together as one family."[85]

Alferov's career took a major turn toward administrative work in the 1980s. Upon Tuchkevich's retirement as director of the LFTI, Alferov was the logical choice to take over the position. An internationally recognized scholar, he commanded great authority in scientific and political circles at home. Having worked at the LFTI under three directors (Komar,

Konstantinov, and Tuchkevich), he knew the history of the institute well. He understood personally and professionally that the LFTI's position as the leading center of solid state physics research in the USSR could be maintained only by dint of hard effort. He had worked closely with Tuchkevich and other laboratory directors in choosing directions of research, finding funding for the research, and working with institutions of higher education to ensure a steady stream of young physicists into the LFTI. He was aware that the system of research and development that had evolved in the postwar USSR had led to a series of serious contradictions. While the USSR faced growing budget pressures as the economy slowed down, its leaders pressed scientists and engineers somehow to make up the slack. The ministerial barriers separating science, education, and production had become an even greater hindrance to the operation of a leading scientific center. And Alferov and other scientific administrators now also had to ensure the health of science in a period of growing political and economic uncertainty under a new Communist Party General Secretary, Mikhail Gorbachev, during the last period of Soviet history—the period of perestroika and glasnost.

4 From Transistors to Heterojunctions[1]

In December of 2000, Zhores Alferov stood at the lectern in the Grand Hall of the Swedish Academy of Sciences in Stockholm to deliver his Nobel lecture. He proudly proclaimed that semiconductor heterostructures were the major foundation of modern solid state physics, and that their applications stretched from CD players at home to solar cells in space. He described his research as grounded in the solid state physics of a "high theoretical, technological, and experimental level" that had been going on at the Leningrad Physical Technical Institute since the 1930s.

When Alferov arrived at the LFTI, in 1953, the institute was suffering through the institutional pains of the ongoing recovery from World War II, a new director, and expansion of research programs far beyond the bread and butter—or silicon and germanium—of solid state physics. The physics of semiconductors, rectification, the theory of contact phenomena, and work on the photoelectric, thermo-electric, galvanomagnetic, electric, optical, and other properties of semiconductors all occupied the physicists, as did important themes connected with the burgeoning nuclear enterprise.[2] Precisely in this research environment, because of a critical mass of young physicists, flexible leadership, an effort to re-emphasize the institute's central programs of solid state physics, and burgeoning interest in transistors, the LFTI housed the development of heterojunctions.

Most modern solid state devices are built on the foundation of the electronic processes in a p-n junction formed by parts of one and the same semiconductor, but with a different type of conductivity. In one part physicists introduce an impurity atom that creates electronic conductivity (of the negative or n type), and in the other an impurity that creates conductivity of the p type (positive). Since the semiconductors differ only in type of conductivity, this kind of p-n junction is called a *homojunction*, in contrast with a *heterojunction* (an interface between two layers or regions of two semiconductors that differ in chemical makeup, preferably with

crystal lattices that match up). The semiconducting materials in a heterojunction have unequal bandgaps, whereas a homojunction has equal bandgaps. The combination of multiple heterojunctions in a device is called
a *heterostructure*, although the terms *heterojunction* and *heterostructure* are
often used interchangeably.[3] Double-heterostructure lasers, with a layer of
a low-bandgap semiconductor sandwiched between two layers of a wide-
bandgap semiconductor, have two significant inherent advantages over
homostructures. Because of the lower conduction-band energy and higher
valence-band energy in the middle layer relative to the wider-gap outer
layers, the electrons and holes are trapped in them. Note that a hole is the
absence of an electron from the otherwise full valance band; it acts in many
ways as the positively charged analogue of an electron. But lower-bandgap
materials also have higher indices of refraction. Thus, in addition to carrier
confinement, the double heterostructure provides optical confinement of
the emitted light to the lower-bandgap region. (See figure 4.1.) The idea of
heterojunctions opened up the possibility of the creation of more effective
devices and the possibility of their miniaturization because the contacts
operate on the atomic scale, perhaps with the thickness of one molecule or
atom. The task of physicists was to determine which semiconductor materials to use.[4]

The twin pillars of today's electronic society are computers and communication systems. Both have grown rapidly since the 1980s, with the
Internet taking a central role in public life of business and government and
in private life. When the Nobel Prize Committee selected Zhores Alferov,
Herbert Kroemer, and Jack Kilby as recipients of the 2000 prize in physics
(Alferov and Kroemer for discovery of the heterojunction, Kilby for the development of integrated circuit), it was acknowledging the centrality of the
heterojunction and the integrated circuit to the revolutions in computing,
communications, and other optoelectric technologies. Alferov, Kroemer,
and Kilby were among the pioneers in the development of those technologies, which led to reliable, low-cost semiconducting lasers, light-emitting
diodes, and other devices.

As Alferov noted in his Nobel lecture, the proposal of p-n junction
semiconductor lasers, the experimental observation of effective radiative
recombination in gallium arsenide (GaAs) p-n structures with a possible
stimulated emission, and the creation of p-n junction lasers and light-
emitting diodes (LEDs) were the seeds from which semiconductor optoelectronics started to grow.[5] Essentially, advances in semiconductor lasers,
diodes, and other devices that yielded higher efficiency, ability to operate

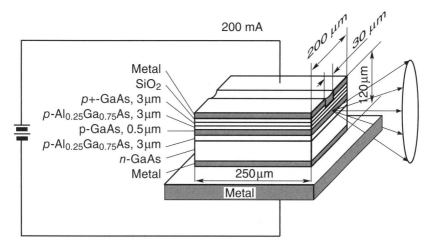

A schematic drawing of the first double-heterostructure injection laser that operated at room temperature. Source: Professor Duncan Tate, Department of Physics, Colby College.

at room temperature, and increased longevity followed a path from the application of simple to complex materials that began in 1930s with copper oxide and continued after World War II with complex combinations of semiconductors. Semiconductors may be elemental materials such as group IV elements silicon and germanium, or compound semiconductors such as those composed of elements from groups III and V of the periodic table (e. g., gallium arsenide and indium phosphide). The heterostructure devices that were first used in semiconductor optoelectronics were made of binary III-V compounds and solid solutions of several III-V compounds. For example, aluminum gallium arsenide is a solid solution of gallium arsenide and aluminum arsenide. Gallium arsenide and aluminum arsenide are unique in that the binary compounds, and the entire range of solid solutions among them, have the same crystal lattice dimensions, so that heterostructures formed by layers of these materials are nearly free of defects. Alferov and his colleagues at the LFTI took advantage of these properties when they developed optoelectronic devices.

Semiconductors are very similar to insulators. The two categories of solids differ primarily in that insulators have larger bandgaps—that is, energies that electrons must acquire to be free to flow. In semiconductors at room temperature, just as in insulators, very few electrons gain enough thermal energy to leap the bandgap, which is necessary for conduction. For this reason, pure semiconductors and insulators, in the absence of applied fields, have roughly similar electrical properties. The smaller bandgaps of semiconductors, however, make it possible to control their electrical properties by means other than temperature.

There are two main types of semiconductor materials: intrinsic ones, in which the semiconducting properties of the material are intrinsic to the material's nature, and extrinsic ones, in which semiconducting properties of the material are manufactured. At first, of course, physicists worked on understanding the materials with intrinsic semiconducting properties. Physicists then learned to modify the intrinsic electrical properties of semiconductors by introducing impurities in a process known as *doping*. Generally, each impurity atom adds one electron or one "hole" that may flow freely. Upon the addition of a sufficiently large proportion of dopants, semiconductors conduct electricity nearly as well as metals.

In certain semiconductors, when electrons fall from the conduction band to the valence band (the energy levels above and below the bandgap) they often emit light. This photoemission process underlies the LED and the semiconductor laser, while semiconductor absorption of light in photodetectors excites electrons from the valence band to the conduction band. Photoemission and photoabsorption have found many commercial applications in fiber optic communications, solar cells, and computing. In addition to doping, scientists and engineers have learned to modify the electrical properties of semiconductors dynamically by applying electric fields. The ability to control conductivity in small and well-defined regions of semiconductor material, statically through doping and dynamically through the application of electric fields, led to the development of the semiconductor devices that are the building blocks of modern electronics, optoelectronics, and integrated circuits.[6]

Today virtually all the semiconductors used in electronic devices are extrinsic. These extrinsic materials, as has been noted, are produced primarily in two ways: by doping and through junction effects that occur when two different materials are joined together. The latter was the focus of Alferov's work. Semiconductor diodes made from silicon are doped to achieve the desired physical and chemical properties. The impurity atoms, for example boron, occupy various positions in the silicon crystal. The matrix of silicon atoms doped with boron atoms is short one electron and hence positive. By the year 2000, scientists and engineers were able to fabricate electronic and optoelectronic devices down to the nanoscale region (that is, generally 100 nanometers or smaller) using high-quality materials grown under well-controlled conditions.

The Rise of the Transistor[7]

Nuclear science was not the only science that experienced rapid growth after the war, with national governments offering a large share of the financing in the hopes of military applications. Rapid postwar growth of the scientific establishment was a sign of researchers' newfound wealth due to their achievements in radar, proximity fuses, nuclear bombs, and so on, and also a reflection of their importance in the Cold War. Both in the United States, with its newly established National Science Foundation, Atomic Energy Commission, Office of Naval Research, and other federal organizations, and in the Soviet Union, with its Ministry of Middle Machine Building (Minsredmash), its Academy of Sciences, and its branch research institutes, the number of researchers and the number and size of institutes grew rapidly, and the efforts of many of those researchers and institutes were tied to or shaped by military R&D.[8] The largesse enabled rapid expansion of all sorts of programs. For example, in the early 1950s, John Bardeen, who had recently left the Bell Telephone Laboratories for the University of Illinois, found a great environment for the expansion of facilities for a research program in solid state physics. The university simultaneously and rapidly expanded its nuclear physics effort and its computer facility.[9]

A number of other American major postwar solid state physics programs gained impetus, among them the Institute for the Study of Metals at University of Chicago (later the James Franck Institute), the solid state laboratory at Bell Labs, MIT's Research Laboratory of Electronics, and the greatly expanded Physics Department at the University of Illinois. AT&T's, RCA's, GE's, and other corporations' laboratories joined academic and industrial centers in refining the basic theory of materials (e.g., germanium and its electronic properties), expanding understandings of conductivity and insulating properties, and advancing knowledge about thermionic emission and photoconductivity. In addition to theoretical advances, practical concerns were paramount; for example, AT&T's research programs enabled the utility to move away from mechanical to electronic switching equipment, and to abandon cathode-ray tubes and vacuum-tube technology for electronic circuits based on solid state devices.[10] In the USSR, the LFTI and Abram Ioffe's Institute of Semiconductors (IPAN) were at the forefront of such efforts.

By the 1830s, Michael Faraday had established that the resistances of many substances decreased with temperature. In the 1870s and the 1880s, Ferdinand Braun discovered rectification (the process of conversion of alternating to direct current). In 1878, Edward Hall demonstrated the deflection of charge carriers in a solid by a magnetic field. In the early 1900s, G. W. Pickard showed that silicon was a good detector material for radio waves. The term *semiconductor* was introduced in the 1910s. In the mid 1920s, semiconductors were successfully developed for use as rectifiers and photo-electric cells; physicists used selenium and copper oxide. By the late 1920s, physicists had determined that trace impurities influenced the behavior of rectifiers both in their resistance to forward flow of current and in their ability to withstand reverse voltages.[11]

In the 1930s and the 1940s, physicists made gradual progress from both empirical and theoretical points of view. Based on the work of A. H. Wilson, physicists advanced the band theory of solids, including semiconductors, and the theory of rectifying contacts: if electrons in a crystalline solid are located in energy bands (instead of single energy levels), then the puzzling differences among the electrical conductivity of metals, semiconductors, and insulators can be explained on a single assumption. If the highest occupied energy band is not quite full, as in the case of metals, then an applied electric field can accelerate electrons in the direction of the field, and an electric current results. If the band is full, then an applied field cannot impart energy to the electrons, and the crystal behaves as an insulator. Semiconductors are insulators with a small forbidden energy gap between the last full band (the valence band) and the first empty one (the conduction band). At elevated temperatures some of the electrons leave the full band and populate the nearby almost empty band. Thus electrons can participate in a conduction process, and the semiconductor becomes increasingly conductive as the temperature is raised (intrinsic conduction). Each electron raised into the conduction band leaves behind an unoccupied state in the valence band. The collective motion of the electrons in the valence band can then be described by the motion of the vacant state, which became known as a *hole*. The hole has properties equivalent to those of a positive charge carrier.[12]

Great interest in and research on semiconductors grew among physicists during World War II, not the least because of their widespread technical application, for example in radars through the magnetron and silicon detectors. Research groups sought to improve the raw material (silicon) and settled on boron as a doping agent. They studied germanium's qualities, and they gained a better understanding of intrinsic and extrinsic semiconductors. They doped germanium with B, Al, Ga, and In to produce p-type (positive) semiconductors, and with N, P, As, Sb, Bn, and Sn to produce n-type (negative) semiconductors.[13] In 1947, the transistor—the "ancestor of microelectronics"—was invented. The invention of the transistor

Zhores Alferov, John Bardeen, Vladimir Tuchkevich, and Nick Holonyak, University of Illinois, Champaign-Urbana, 1974.

contributed to a revolution in science and technology, and then to a post-industrial revolution of solid state circuits, computers, mobile phones, and so on.

Not long after the development of the transistor, ideas for controlling the holes and electrons in them started to emerge in what has become known as *bandgap engineering*. In his 1948 transistor patent application, William Shockley mentioned the idea of using a wider-gap semiconductor for the emitter than for the rest of the transistor. Soon transistorized hearing aids and radios would be available.

Vladimir Tuchkevich understood the significance of the transistor immediately and proceeded to organize a sector of semiconductor electronics, to which he attracted Alferov and other young physicists. In the 1950s, the group energetically pursued one device after another and one application after another. Their research theme was ploskost', meaning plane. Work on planar transistors and diodes quickly became the basis for research on semiconductor electronics in Tuchkevich's laboratory and throughout the LFTI. Germanium and silicon p-n junctions were the foundation of powerful semiconductor devices that had applications in a variety of fields, including transportation (electrification of railroads), long-distance electrical transmission, control (some of the first devices were used in atomic submarines), welding, and electrolysis.

Because it was small, reliable, and functional at room temperature, the transistor facilitated the designing and building of complex electronic systems for communications, data processing, guidance, and control. Yet relative to today's solid state circuits, the increasingly complex systems that were built for televisions, computers, spacecraft, and other applications were heavy and bulky. Integrated circuits based on homojunctions and then on heterojunctions facilitated miniaturization. They not only replaced large functional groups of discrete components, they also avoided "the problem of failures in the connections between components."[14]

All over the world, a large number of researchers turned their attention to the development of new methods for the purification and production of artificial semiconductors—monocrystals of germanium and later silicon. Every success led to new efforts to perfect transistors. The replacement of silicon with germanium made it practicable to raise the working temperature of instruments to room temperature and to use high voltages. But in the early 1960s physicists noticed a problem that became greater with the appearance of the first semiconductor injection lasers. Electron-hole junctions in a single semiconductor, i.e. homojunctions, were limited in performance, and semiconductor lasers could function only at very low temperatures.[15] Kroemer, Alferov, and others stepped in here, with Jack Kilby focusing on integrated circuits.[16]

Kilby joined Texas Instruments in the summer of 1958. After considering the "alternate miniaturization schemes that were around at that time," he decided to use semiconductors "not just for transistors and diodes, but also for resistors and capacitors, putting everything on the same substrate." In a few months he succeeded in producing integrated circuits of these components, first using germanium and then silicon. He had to solder all the components together by hand, a process that was labor intensive and error prone. Some of the components were interconnected by virtue of their arrangement on the germanium; others had to be connected with gold wire.[17] Robert Noyce of the Fairchild Semiconductor Corporation then drew on work done by his colleague Jean Hoerni and by John Moll and Carl Frosch at Bell Labs to interconnect flat transistors by depositing metal over etched holes in the oxide. Noyce thus became a co-inventor, with Kilby, of the integrated circuit. Gordon Moore, who in 1968 would leave Fairchild with Noyce to found the Intel Corporation, observed in 1965 what has come to be known as Moore's Law: the exponential growth of semiconductor devices in capacity, speed and quality of materials every eighteen months or so.

Integrated circuits developed along two lines: silicon monolithic circuits and thin-film circuits. The integrated circuit grew out of transistors and developed rapidly owing to the favorable properties of silicon, on which the epitaxial planar process is based. Additionally, the oxide of silicon could be grown on the surface to a determined thickness to serve as a passivation layer (that is, to make it passive in relation to another material), as an insulator, or as a dielectric. Since many transistors could be made on a single wafer of silicon, additional processing to interconnect them was not difficult. In larger, more complex integrated circuits, interconnections were required in more than one plane; here thin-film deposition techniques were required for the insulation between crossovers. Silicon was far from a perfect insulator, so "parasitic losses and spurious coupling between elements on the same substrate could seriously limit circuit performance."[18]

In 1957, Herbert Kroemer, then at the RCA Laboratories in Princeton, returned to an earlier idea of his to vary the energy gap in semiconductors.[19] He realized that a spatially varying energy gap, which results in variations in the semiconductor's conduction and valence bands, creates quasi-electric fields that act differently on electrons in the conduction band and holes in the valence band. The resulting forces on electrons and holes could even act in the same direction. Kroemer understood that this effect should make it possible to create devices that are fundamentally impossible to achieve in homostructures. He told *Physics Today*: "While the examples I could come up with—transistors with a graded bandgap in the base, or with a wide-gap emitter—represented improvements . . . they did not fully exploit the new principle." Kroemer did not, however, develop any heterostructure laser until well after Zhores Alferov, Izuo Hayashi, and Morton Panish demonstrated heterostructure lasers capable of continuous wave (CW) operation at room temperature. His major contribution was the initial suggestion, which went unnoticed by most physicists and engineers.

Kroemer received a PhD in theoretical physics in 1952 from the University of Göttingen. His dissertation was on hot-electron effects in a transistor. He then took a position as a theoretician in semiconductor research in the telecommunications laboratory of the German Postal Service, where he focused on why emerging junction transistors did not compare in speed to the earlier point transistors. According to the physicist and historian Manfred Thumm, this led Kroemer to ask "How can an electric field be built into the base region of a junction transistor?" He determined that the best way to do this was "not by using a single semiconductor, but a graded base region that started with one material and ended in another with a continuous transition between the two."[20] At this point, in the absence of data, there was little progress.

In the mid 1950s, Kroemer showed the advantages of heterostructures in various semiconductor devices by incorporating heterostructures into them. In 1957, while at RCA's David Sarnoff Research Center in Princeton, he developed the idea of the heterostructure bipolar transistor, basing it on the idea that "composition gradients acted as forces on electrons." He returned to Germany in 1957 to head a semiconductor group at the Philips Research Laboratory in Hamburg, where he pushed for work on GaAs. He then moved, in 1959, to Varian Associates in Palo Alto, where he remained until 1966, and where he developed the idea for a double-heterostructure laser that would operate continuously at room temperature, writing a seminal paper on the topic. Kroemer later recalled:

I wrote up a paper describing the DH idea, along with a patent application. The paper was submitted to *Applied Physics Letters*, where it was rejected. I was urged not to fight the rejection, but to submit the paper to the *Proceedings of the IEEE* instead, where it was published, but largely ignored. The patent was issued in 1967. It is probably a better paper than the *Proc. IEEE* letter. It expired in 1985.[21]

Kroemer filed and received a patent for his invention, but was discouraged from work on light-emitting semiconductors because other specialists could not see applications. In part this was because of the requirement that new materials and processes be developed. The lasers required the development of modern epitaxial growth technology before they could become mainstream technologies.[22]

In the early 1960s, the semiconductor community began to turn its attention to developing a laser. In a lamp, the spectrum of color should be wide, as in sunlight. If a lamp provides a narrow spectrum, you see only that color. Normally photons behave independently, probabilistically. In a laser, all the photons "march together," like soldiers.[23] Hence, a laser is any of several devices that emit highly amplified and coherent radiation of one or more discrete frequencies. One of the most common lasers makes use of atoms in a metastable energy state; as they decay to a lower energy level, they stimulate other atoms to decay, which results in a cascade of emitted radiation. Semiconductor lasers are very small and can be switched on and off with high frequency. They can help transmit of a lot of information, as the lasers used to read the information on CDs and DVDs and those used in fiber optic communications systems do.

Semiconductor lasers owe their genesis largely to work done by physicists in the United States and the Soviet Union during the Cold War—work that focused on their potential military applications. But lasers attracted the attention of scientists in hundreds of laboratories around the world "not only due to their wide applications—from compact discs to fiber optical communications—but also to their unique, by diversity and scale, physical phenomena."

The bandgaps of silicon and germanium made those elements unlikely candidates for lasers, so the focus was on III-V semiconductors such as GaAs. The goal of a laser was attained in 1962 by several groups. Robert Hall and co-workers at General Electric in Schenectady, Marshall Nathan and colleagues at IBM, and Robert Rediker and co-workers at MIT's Lincoln Laboratory used GaAs; Nick Holonyak and Sam Bevacqua at General Electric in Syracuse built a visible laser using the alloy GaAsP. But these lasers provided did not achieve CW operation at room temperature, and they were short-lived. High threshold currents and low temperatures would be necessary.

According to Morton Panish, early GaAs homostructure lasers evolved into a variety of heterostructure lasers with commercial applications. In the period 1958–1961, Nikolai Basov and Bentsion Vul of the Physics Institute of the Academy in Moscow, W. S. Boyle and D. G. Thomas of Bell Labs, and other physicists suggested that semiconductors might be used as laser materials. Next, the suggestion arose to use III-V compounds, for example

GaAs and GaSb, for heterostructure injection lasers. Alferov and his colleagues, Hans Rupprecht at IBM, and (somewhat later) Hayashi and Panish recognized the importance of using lattice-matched semiconductors and so arrived at the naturally lattice-matched system GaAs-Al$_x$Ga$_{1-x}$As. Here the subscript x can be any value from 0 to 1 for the complete range of solid solutions between GaAs and AlAs. In this important work of the late 1960s and the early 1970s, x was usually about 0.3. In subsequent years, other materials were used to meet important wavelength regions not attainable in the GaAs-AlGaAs system. Notably, GaAs-GaInAsP heterostructure lasers were used to achieve the longer-wavelength emissions needed for optical fibers.[24]

Transitions in semiconductors were over a much broader range of energies than the characteristically sharp transitions in conventional lasers. The idea of obtaining laser action in a semiconductor was discussed at IBM in 1961, and awareness of the possibility of efficient electroluminescence in GaAs at 77°K was evident in March of 1962.[25] In July of 1962, physicists discussed a variety of issues at the Solid State Device Research Conference in Durham, New Hampshire. The participants included Nick Holonyak, who was studying tunneling in the crystalline solid solution GaP$_x$As$_{1-x}$ in a GE laboratory. After the conference, Holonyak prepared p-n junction diodes by diffusing Zn into n-type GaP$_x$As$_{1-x}$ layers on wafers that he grew by chemical-vapor deposition.[26] Other papers published at this time also reported injection lasing.

At the 1963 Solid State Device Research Conference, held in East Lansing, Michigan, several physicists delivered reports on diode lasers using different materials, on their efficiency, and on temperature threshold. Early in 1962, D. N. Nasledov et al. had reported slight narrowing of the electroluminescent spectrum of the GaAs diode at 77°K.[27] Yet the threshold current for lasing was very high at room temperature; most studies were conducted in liquid nitrogen at 77°K or lower, and enthusiasm waned. In 1963, Herbert Kroemer suggested that improved junction lasers could be achieved with a structure in which a layer of semiconductor with a relatively narrow energy gap is sandwiched between two layers of a wider energy gap semiconductor (the heterojunction). In a 1963 Soviet patent application, Zhores Alferov and Rudolf Kazarinov proposed double-heterostructure lasers. Kroemer applied for a US patent several months later.[28] Hence, on learning of the large optical and electrical losses, Alferov and Kroemer, neither of whom had any background in lasers, independently proposed the solution: a double heterostructure. At that time neither of them proposed lattice matching, so their patents were useless in practice. Kroemer did not follow up, and it was some years before Alferov latched onto the idea of lattice matching (around the same time as Hayashi and Panish). Nor was Alferov's patent made public.

Double heterostructures of AlGaAs-GaAs and later of other substances made long life, room-temperature operation, and high power possible. Double-heterostructure lasers, with a low-bandgap material sandwiched between two wide-bandgap semiconductors, have two significant inherent advantages over homostructures. Because of the lower conduction-band energy and higher valence-band energy in the middle layer, the electrons and holes are trapped there. But lower-bandgap materials also have higher indices of refraction. Thus, in addition to carrier confinement, the double heterostructure provides optical confinement of the emitted light to the lower-bandgap region. Kroemer proposed to use the double heterostructure for carrier confinement in the active region. He suggested that "laser action should be obtainable in many of the indirect gap semiconductors and improved in the direct gap ones, if it is possible to supply them with a pair of heterojunction injectors."[29] According to Panish, Kroemer was incorrect in this suggestion.[30] Heterostructures devices, made primarily of GaAs-GaAlAs and GaAs-InGaP but increasingly of Si-Ge alloys, are used in cell phones and satellite receivers because they can operate at very high frequencies and with low noise. Heterostructures have emerged as the basic building block of semiconductor devices. "Over 99 percent of the semiconductor research today involves heterostructures," says Alferov. The title of a talk Kroemer gave in 1980 has come true: "Heterostructures for Everything."[31]

In the early 1960s, Alferov organized an effort at the LFTI to explore heterostructure applications, and many American labs continued to work on semiconductor devices. However, Kroemer, working at Varian Associates in Silicon Valley, was refused support for work on double heterostructures, turned his attention to the Gunn Effect, then returned to heterostructures in the early 1970s, focusing on heterojunction bipolar transistors.

Kroemer's work on heterojunction injection lasers may have shown the direction for the development of the semiconductor lasers used in CD players, fiber optics, and other applications. But by the mid 1970s a number of Japanese and American laboratories were already following the lead of heterostructure laser work done by Hayashi and Panish and by Alferov's group. In addition, Panish and others developed the first high-speed heterostructure transistors. (Kroemer also did pioneering experiments in molecular beam epitaxy, concentrating from the outset on applying the technology to untried new materials systems, such as GaP and GaAs on silicon substrate. In the late 1990s, he turned again to theoretical work. His work in heterostructure-based transistors has furthered the development of the cell phone and other wireless communications technologies.[32])

It was difficult to commence work on new materials when few physicists believed success was imminent. One reason for this is that theoretical understanding moved ahead more rapidly than experimental verification. In 1966, Alferov and others predicted the density of injected carriers could exceed the carrier density in the wide-gap emitter by several orders of magnitude (the "superjunction" effect).[33] Others were skeptical of the possibility of creating the "ideal heterojunction" with a defect-free interface and with theoretical injection properties. For example, R. L. Anderson's pioneering study of the first lattice-matched epitaxially grown single-crystal heterojunctions of Ge-GaAs did not provide proof of injection.[34]

But Alferov had the support of the LFTI's directors, especially Boris Gaev, the head of the institute's operations department. How did Gaev assist? Gaev, a naval engineer in the 1930s, joined Konstantinov in the study of the separation of ^{6}Lithium for the hydrogen bomb, for which the group won a Lenin Prize. While some physicists were skeptical of Alferov's small heterostructure group, Gaev made sure he had an office, rooms, and assistants. When Alferov asked him why he could rely on Gaev's support, Gaev replied "I don't understand what, precisely, you are doing, but I understand what you did earlier and I am certain this won't turn out to be nonsense."[35] Alferov told me several reasons for his turning to this problem: he knew transistors inside and out, he understood semiconductor materials fluently, he had had enough of rectifiers, and his recent divorce had left him somewhat adrift. "Lasers excited me," he said. Having read about Anderson's work on heterostructures at IBM, he decided to design heterostructure lasers. He and others made rapid progress, achieving adequate room-temperature electronic and optical parameters by 1965 and superinjection shortly afterward.

Tuchkevich recalled: "When we [the Alferov group at the LFTI] began this research [on heterostructures] we did not think about success. We were young and simply plunged into the problem, seeking an 'ideal pair' [of semiconductor materials]." A heterojunction was a pair, and the choice of such a pair was a complex problem. The group developed dozens of different contacts. Not all heterojunctions were ideal. Eventually they tried gallium arsenide (which had been well studied) and aluminum arsenide. "In order to begin research on this pair, we had to overcome a definite psychological barrier," Tuchkevich acknowledged. "Although aluminum arsenide was synthesized already in 1928, its characteristics remained little studied for one reason—it was highly corrosive and unstable, and [it] decomposes literally before your eyes in a humid environment."[36] Creating heterojunctions of these two materials seemed highly unlikely.

Alferov and others had first begun to explore junctions of III-IV materials under Nina Aleksandrovna Goriunova, who had then died suddenly, very young. One of her students, Dmitrii Tretiakov, who worked fairly closely with Alferov, mentioned that he had fortuitously discovered that unstable arsenide of aluminum was absolutely stable in the compound aluminum gallium arsenic. In 1966, using this material, physicists in Alferov's laboratory created the first lasers based on heterostructures. Those lasers worked at room temperature, but they did not operate continuously and they had a low threshold current. Two years later, Alferov and his co-workers Dmitrii Tretiakov, Dmitrii Garbuzov, Vizcheslav Andreev, and Vladimir Korol'kov built the first semiconductor heterolaser that worked constantly (not in pulses) at room temperature, with a significant theoretical contribution from Rudolf Kazarinov.[37] Alferov et al. reported these continuous-wave room-temperature semiconductor lasers about a month ahead of Izuo Hayashi and Morton Panish of Bell Labs.[38]

LFTI physicists eventually determined they could overcome surface corrosion by introducing atoms of gallium into aluminum arsenide. They took the characteristics of the new heterojunction, changed the parameters of layers, looked at the heterojunction under an electron microscope, and moved on. A small group worked on this, each member adding something hitherto undiscovered. "We believed in each other," said Alferov's colleague Vladimir Korolkov. "And this isn't simply comradely trust—it was trust of a comrade-in-arms. Each of us studied to exactitude our part of the work. And, having passed along the researched item to the next, each was certain that the work would be continued with the same great care. And if we had worked alone, imagine . . . there wouldn't be those almost 80 published journal articles which have now been published in *Fundamental Research of Heterostructures in Semiconductors*." As for applications, Korolkov said, "I saw new, more perfect transformers of solar energy into electric . . . [that] work at such high temperatures that solar batteries give out."[39] The physicists anticipated dozens of applications.

To put it more finely, the major problem in the development of heterojunctions was the selection of materials to ensure matching of the lattice constants of the semiconductors. GaAs-GaAlAs eventually emerged as the pairing of choice. Researchers determined the close values of their lattice constants and noted that, whereas AlAs will oxidize in air, the alloy $Ga_xAl_{1-x}As$ is chemically stable for low to moderate Al concentrations. Heterostructure research picked up steam in 1967 when Alferov in his laboratory and Hans Rupprecht and Jerry Woodall at IBM independently reported the growth of $Al_xGa_{1-x}As$ on a GaAs substrate by liquid-phase epitaxy; Herbert Nelson at RCA first reported this process for GaAs on GaAs.[40] Panish and Hayashi first tested single heterostructures to determine whether the seemingly ideal heterojunction really existed. Having demonstrated that even the very simple single heterostructure lased had dramatically lower lasing threshold current than homostructure lasers, they proceeded to double heterostructures. At first, Alferov's group focused on double heterostructures. Over the next few years, his group achieved efficient carrier injection, the predicted electrical and optical confinement, and low-threshold pulsed room-temperature lasing in such devices. After the CW lasing demonstrations by Alferov and by Panish and Hayashi, research progressed to the study of other lattice-matching semiconductor combinations in order to cover a broader area of the optical spectrum. Alferov et al. reported in 1970 that "various lattice-matched heterojunctions based on quaternary III-V solid solutions were possible which permitted independent variation between lattice constant and bandgap."[41]

The important thing, once again, was to find a suitable set of semiconductors to prepare the heterojunction based on the principle of a lattice match between the two different semiconductors. At the LFTI and in other laboratories, researchers discovered that $Al_xGa_{1-x}As$ solid solutions were chemically stable and suitable for the task of producing double heterostructures and devices. From this point on, this compound was adopted fairly rapidly around the world.

Panish and Hayashi, who were new to this area, joined in the effort to create injection lasers capable of continuous operation at room temperature. When they set out to create heterostructure lasers, in late 1966 or early 1967, neither of them knew much about lasers, and Panish was new to semiconductors. They were not initially aware of Alferov's work, although Panish believes they "should have been": "We had already done the single heterostructure by the time we became aware of Alferov. Had we been initially aware we would certainly have skipped the single heterostructure laser as we were pretty sure that we would eventually have to go to a sandwich structure. Our failure to do an adequate literature search still haunts me. We missed Kroemer's early work also."[42]

In January of 1969, when Panish and Hayashi submitted a paper in which they proposed what they thought was the first heterostructure laser, the reviewers of their article pointed out that the 1963 paper of Kroemer, the paper of Kressel and Nelson, and work by Alferov et al. indicated previous work on such lasers.[43] In 1970, at a conference in Germany, Alferov invited Panish and Hayashi to visit his laboratory in Leningrad. Alferov had visited them at Bell Labs in September of 1969, and it was clear to him that his laboratory was ahead. During their September 1970 visit to Alferov's laboratory, they found Alferov generous, open, and friendly as they toured his laboratory, but neither party acknowledged fully how far they had come in developing the new laser devices. "In September 1969," Panish and Hayashi wrote, "Zhores Alferov . . . visited our laboratory. We realized he was getting [excellent parameters for a double heterojunction]. We had not realized the competition was so close and redoubled our efforts. . . . Room temperature CW operation was reported in May 1970. . . ."[44] The Alferov group published a paper, also in 1970, that discussed the realization of CW lasing.[45] Independently (as Alferov explained in his Nobel speech), in a paper submitted only a month later, Hayashi and Panish reported CW operation for double-heterostructure lasers.[46] All this work produced a storm of interest, with research expanding rapidly from the LFTI, IBM, Bell Labs, and RCA to universities and laboratories throughout the United States, the Soviet Union, the United Kingdom, Japan, Brazil, and Poland. Clearly, scientific competition was a greater motivation than competition between the Soviet Union and the United States.[47]

Alferov and colleagues worked on GaAs-Al$_x$Ga$_{1-x}$As and reported studies of injection across GaAs-Al$_x$Ga$_{1-x}$As p-n heterojunctions in 1968, and in 1969 they published a paper concerning a sandwich of p-GaAs between n- and p-Al$_x$Ga$_{1-x}$As layers grown onto a GaAs substrate. In September of 1970, they published in *Fizika i Tekhnika Poluprovodnikov* a paper (submitted in May) in which they claimed to have achieved CW lasing with a double-heterostructure laser at 300°K. Panish and Hayashi had reported similar work in June.[48]

Light-Emitting Diodes

The two semiconductor sources of light are lasers and light-emitting diodes. Lasers have many applications, from medicine to entertainment. LEDs are now used in billions of devices, including indicator lamps, clocks, home theater equipment, timers in stoves and microwave ovens, and traffic lights. Scientists have developed LEDs that produce white, blue, red, orange, and green light.

The birth of LEDs may be traceable to the early 1920s and the laboratory of the Leningrad physicist Oleg Vladimirovich Losev. The British engineer Henry Joseph Round had already noted that crystal detectors emitted light when a current was passed through them. But the yellow light was too dim to be of practical use, and difficulties in working with silicon carbide caused Round to abandon the research. Losev, an amateur radio enthusiast who worked on crystal detectors, noticed the effect in 1922, but he died during the Leningrad blockade, and much of his work was lost. In the late 1920s physicists used materials made from zinc sulfide doped with copper (ZnS:Cu), but these materials emitted only a low level of light. George Destriau published a report on the emission of light by ZnS powder in 1936; physicists credit him with first using the term *electroluminescence*.[49]

The first LEDs were developed in the early 1960s under the leadership of Nick Holonyak and others. Scientists and engineers encountered significant difficulties in manufacturing LEDs that operated at room temperature; the early ones had to sit in liquid nitrogen while operating. The first visible (red light) commercial LEDs were used in sensing and photoelectric applications. Over the next decade, physicists employed a series of substrates (for example gallium arsenide phosphide [GaAsP], then gallium phosphide [GaP]) that increased efficiency, produced brighter red lights, and allowed orange, pale green, and yellow light to be produced, the latter using dual GaP chips (one in red and one in green). Yellow LEDs were made in Soviet Union using silicon carbide at around this time, but they were very inefficient. The use of gallium aluminum arsenide phosphide (GaAlAsP) LEDs in the early to mid 1980s brought the first generation of super-bright LEDs, first in red, then yellow, and finally green. By the early 1990s, ultra-bright LEDs that used indium gallium aluminum phosphide (InGaAlP) to produce orange-red, orange, yellow, and green light had become available.

Mass Production of High-Quality Heterojunctions: Epitaxy

Over the past half century, scientists and engineers have developed a series of techniques for producing super-thin semiconductor materials. In epitaxy, a thin layer (0.5–20 microns) of single-crystal material is deposited over a single-crystal substrate by means of a chemical vapor. In the postwar years, thin films found industrial use as anti-reflection coatings, front-surface mirrors, interference filters, and decorative coatings, in the manufacture of cathode-ray tubes, and then in electronic circuits. Michael Faraday probably produced the first evaporated thin films in the 1850s. Some years later, R. Nahrwold and then August Kundt deposited thin metal films in a vacuum.[50] The three ways of depositing a thin film are molecular beam epitaxy (MBE), metal organic chemical vapor deposition (MOCVD), and liquid phase epitaxy (LPE).[51] Early on, in semiconductors, the deposited film was often the same material as the substrate (homoepitaxy)—for example, silicon deposition over a silicon substrate. By growing a lightly doped epitaxial layer over a heavily doped silicon substrate, specialists achieved a higher breakdown voltage across the collector-substrate junction while maintaining low collector resistance. Lower collector resistance allowed a higher operating speed with the same current.[52]

To pursue applications, physicists found it necessary to produce heterojunctions in high quantity in pure environments. "Ribbons" of semiconductor materials were the foundation of this technology, Alferov and his colleagues believed. They powered various instruments, filters, lasers, and prisms. Although the industrial methods used to produce them were not yet entirely cost-effective, they soon would be, especially as their size and thickness decreased with simultaneous increases in efficiency. In the early 1970s, crystals of semiconductor materials 30–50 mm wide and 200–300 mm long were produced. These were too big, and had to be cut by diamond disks into plates 0.2–0.5 mm thick. The process ruined some of the crystals, required further processing, and resulted in waste of material. A. V. Stepanov, a physicist at the LFTI, suggested a new method: with a special device, a small volume of liquid could be added to the necessary form, and in crystallization the form would be preserved. Several years of close scientific contact between the LFTI, the All Union Research Institute of Electrothermal Instruments, and an enterprise under the jurisdiction of the Ministry of Non-Ferrous Metallurgy (MinSvetMetal) were required to introduce the process.[53] It should be kept in mind that thousands of people worked on these problems over the decades, and that the Soviet efforts were not outstanding relative to German, Japanese, and American efforts and may indeed have lagged substantially in time and quality.

The evaporation of thin films with good characteristics requires a superclean operating environment to avoid any interference with the process of film formation. This was achieved by using a high vacuum to minimize the interaction between residual gases and the surfaces of growing films. Single-crystal films gained great importance in connection with the rapid development of solid state circuitry, but avoiding such defects as dislocations, vacancies, and stacking faults was a challenge. In addition, poorly controlled cooling might result in mechanical stress and deformation. Epitaxy evolved in the effort to produce single-crystalline films without all these problems.[54] The advantages of thin-film elements over silicon were considerable: all elements were deposited on an insulating substrate, such as glass or ceramic, which provided electrical isolation between them without any special techniques, and the manufacturing required fewer oxidation, photoetching, diffusion, and metal-deposition cycles or processes. The first advanced all-thin-film integrated circuit, produced in 1965, contained 360 thin-film transistors, 180 diodes, 360 resistors, and 180 capacitors.[55]

Specialists around the world, including at the LFTI, developed the process of vacuum deposition of a material onto a substrate. An atom or a molecule of vapor arrives on the substrate. It may re-evaporate, migrate over the surface, and rest in isolation, or it may migrate to an atom or a group of atoms to which it becomes bonded. Problems in manufacture could arise if there were impurities, and the ratio of the deposition rate to the pressure determines the proportion of trapped gas. Until the late 1960s, the materials most commonly used in thin-film transistor fabrication were two II-VI compounds: cadmium sulfide (CdS) and cadmium selenide (CdSe).[56]

Metalorganic chemical deposition (MOCVD) involves passing metal oxides across a work piece in an inert gas to deposit a layer of metal oxide on the surface. The base wafer—silicon or germanium—must be heated on a graphite susceptor (a material that absorbs electromagnetic energy well) to a very high temperature in order to accept the deposition. A semiconductor material is produced by doping the silicon wafer with aluminum (to form a p-type semiconductor) or with boron (for an n-type semiconductor). At the LFTI, researchers began to develop MBE and MOCVD methods of growing III-V heterostructures only in the late 1970s. They stimulated the design and construction of the first Soviet MBE machine in the electronics industry. The Scientific Instruments Production Organization of the Academy of Sciences in Leningrad also commenced work on MBE machines in concert with the LFTI. The physicists developed an MOCVD system at the LFTI, and a few years later worked with the Swedish firm Epiquip to produce machines that are still in use at the LFTI today.[57]

Today the MOCVD process ensures good material distribution and reproduces fine details with reliable quality. At first, in the 1970s, it resulted in low-quality heterojunctions because of low-quality source materials. But in the 1980s, two state-of-the-art factories, one in Novosibirsk and one in Leningrad, commenced operation and turned out the materials at 700–800°C and at a 10^{-9} vacuum. The MOCVD process, which is more environmentally sound than the MBE because the contents can be controlled, is preferred in the United States; in Russia, MBE is preferred because of its lower cost.[58] For the first MBE devices, scientists had to produce their own gases, which was both inefficient and difficult. Now they buy them in industrially produced canisters. The Aixtron machine, a well-respected MBE machine manufactured in Aachen, Germany, uses a nitrogen and argon environment.[59] Western physicists gained an advantage because they could buy substrate crystals from the Siemens Corporation and other sources, and they learned to grow substrate crystals earlier and in cleaner environments than the Soviet physicists, who were handicapped by a scientific instrument building enterprise that lagged behind that of their Western colleagues.[60]

Microapplications in the Macroworld

Specialists quickly discovered wide and significant applications in a variety of fields. Optoelectronic devices include detectors and emitters with numerous applications in optical communications, terahertz applications, infrared and long-wavelength infrared detectors, and multi-junction solar cells. One of the mechanisms used to generate light from a semiconductor is the radiative recombination of electrons and holes across the fundamental bandgap, which gives rise to photon emission. Soon after the invention of the laser, p-n junction GaAs lasers were demonstrated with an emission outside of the energy spectrum visible to the human eye. Subsequently, researchers focused on the development and production of emitters in the visible, ultraviolet, and infrared spectral regions (for example, III-nitride semiconductor materials led to the production of blue and green LEDs and diode lasers).[61] Heterostructures have a near universal importance in opto- and consumer electronics and in basic research. For example, they are central to modern transistors. In high electron-mobility transistors (HEMTs), these heterostructures can operate at very high frequencies and with low noise. Such devices, primarily of GaAs-GaAlAs and GaAs-InGaP but increasingly of Si-Ge alloys, can be found in cell phones and satellite receivers.[62] As for basic research, Horst Stormer and Daniel Tsui first observed the fractional quantum Hall effect in 2DEG (two-dimensional electron gas) heterostructures, while ballistic transport and quantum-dot "artificial atoms" occupy a number of scientists.

A crucial application that can draw on advances in heterostructures is solar power. Alferov considers this one of the most important areas on which to focus. In view of the rapid growth of consumption, the world's ever-increasing population, and the externalities of pollution, greenhouse effect, and radioactive waste from nuclear reactors, Alferov believes that solar power may be the best alternative. Some individuals even sanguinely refer to the possibility that global warming might enable Siberia to become a productive agricultural region. Alferov grants that an increase of a few degrees in Siberia's average annual temperature might be beneficial, but believes that overall global warming will contribute to "irreversible and potentially dangerous changes" with great consequences. On the other hand, solar power seems safe and virtually limitless. Mean solar power received by earth far exceeds the amount that might be used "if highly efficient methods of solar power conversion are used." Opponents of solar power have argued that large-scale solar energy conversion requires very large areas, that the conversion of solar power is expensive, and that the technology requires unrealistically high expenditures in manpower and material. At 10 percent efficiency, projecting to demands for electrical energy in 2020, an area of solar receivers of 12,500 km^2 would be required. But Alferov argues that increases in efficiency make this a fruitful area for study.

Regarding solar energy, in his research program Alferov once again drew upon the tradition of Ioffe Institute. As he points out, in the 1930s at that institute B. T. Kolomiets and Iu. P. Maslakhovets produced thallium sulfide photocells with a then record high efficiency of about 1 percent. Abram Ioffe, who in 1934 published in a paper titled "Semiconductors: A New Material for Electrical Engineering," spent the last 20 years of his life investigating the thermoelectric applications of semiconductors. And indeed engineers had begun to find applications. During World War II they developed thermoionic generators as power supplies for radio transmitters. By 1954, Chapin, Person, and Fuller produced silicon p-n junction photocells with efficiency of about 6 percent. The space age brought forth further practical applications, with photocells powering virtually every spacecraft since Sputnik 3 and Vanguard 1, although small nuclear reactors and thermal nuclear devices have been used in some satellites.[63]

Alferov takes exception with those who argue that generating electricity by solar power is reasonable only in areas distant from major grids, as the cost very high. In 1975 the total power of all solar cells in the United States was only 100 kilowatts, at a cost of more than $20,000 per peak kilowatt. But with semiconductor-grade silicon, new materials, and novel types of converters, scientists achieved laboratory efficiencies of 18 percent and commercial efficiencies of 12–14 percent, while photocells based on AlGaAs heterostructures provided efficiencies of 20–25 percent at sunlight concentrations. The cost of solar-generated electricity was reduced two- to threefold by 1985, and the goal of making it only two or three times more expensive than conventional sources of power by 2000 was nearly achieved.[64] The cost of solar energy dropped tenfold between 1985 and 2005.[65] With the price of a barrel of oil having approached $140 in 2008 before dropping under $80, and with nuclear reactors having capital construction costs of $6 billion per 1,000 megawatts of installed power (not including the extensive direct and indirect governmental subsidies to the industry), solar power must have a place in the early twenty-first century, Alferov believes.

Alferov has considered which devices and materials are best to use (p-n junctions, heterojunctions, or Schottky barriers? inexpensive silicon or expensive gallium arsenide?). He points out that the differences in efficiencies between silicon and gallium arsenide are not really large, that the technology for producing semiconductor grade silicon is well developed, and that silicon is well understood and is the second most abundant element on earth after oxygen. Photocells of single-crystal silicon can made of relatively inexpensive materials. New methods for growing ribbon-shaped silicon and automated epitaxial and ion implantation methods of producing p-n-n+ structures lead to optimism concerning reduction of manufacturing costs. Alferov and others also believe it will be possible to develop solar cells of inexpensive semiconductor materials (polycrystalline silicon, Cu_2S-CdS heterostructures) or complex heterostructures based on A3-B5 compounds—primarily AlGaAs heterostructures.[66] Nanotechnologies are currently Alferov's major focus.

5 Perestroika and Politics

Alferov as Scientific Administrator

During the 1980s, Zhores Alferov rose to the top of the scientific establishment. He became a full member of the Academy of Sciences, one of its eleven vice presidents, the director of the Leningrad Physical Technical Institute, and the chairman of the Leningrad Scientific Center. He remained active in research in solid state physics, although administrative responsibilities slowed his work. Travel between the LFTI near Sosnovka Park and the presidium building of the Academy on the Neva River, and between Leningrad and Moscow, even on the fast, comfortable overnight train, the Red Arrow, occupied more and more of his time.

Alferov's ascent of the administrative ladder coincided roughly with Mikhail Gorbachev's rise to power. Gorbachev embarked on reforming the Soviet system within weeks of becoming Secretary of the Communist Party in March of 1985, pushing perestroika (meaning, variously, major economic and political reforms, revolution, or rebuilding) and glasnost (meaning openness to public discussion of previously taboo subjects—corruption and privilege in the Communist Party, the true nature and human costs of Stalinism, and so on). Alferov generally saw Gorbachev's leadership as a great opportunity both for the nation and for science. In his own reform activities, Alferov focused more on science than on politics, but he was deeply excited about the prospects of perestroika generally.

Alferov wished to improve the performance of science, in particular by fostering closer connections between institutions of higher education and research institutes. In spite of the importance of science to the USSR's leaders, Soviet science failed to perform at a level commensurate with the resources bestowed upon it. Alferov found it increasingly bureaucratic and dominated by elder statesmen unwilling to make changes in the way basic research was organized and funded; in 1985, at the age of 55, he did not consider himself one of these elder statesmen. As for science, he did not see the American university system—with its seemingly healthy combination of teaching and research, experience and youth, rigorous national competition, and decentralization of programs—as a panacea for Soviet problems. But he recognized that some features of Western science might be employed in the Soviet system to good effect. Any reforms introduced into Soviet science by Alferov and others had little impact, however, largely because of the short time frame (the USSR collapsed in 1991, and with it the scientific establishment) but also because of the ossification and resistance of the scientific bureaucracy and the rapid deterioration of the economy (the latter provoked, ironically, by Gorbachev's modernization efforts).

Fundamental Science and the Gorbachev Reforms[1]

Mikhail Sergeievich Gorbachev was named General Secretary of the Communist Party on March 18, 1985, the day after the death of his predecessor, Konstantin Chernenko. Chernenko's brief rule epitomized the "muddling through" that characterized Soviet politics and economics from the last years of the Brezhnev regime.[2] Soviet leaders faced a series of intractable problems but were unable and unwilling to act in novel ways to alter the status quo. Economic growth had declined to 2 percent per annum, down from the 6 percent of the 1950s and the early 1960s. Expenditures on health care and the consumer sector dwindled. Brezhnev's highly touted Food Program (1982) barely made a difference in improving agricultural production. Young people left the countryside in droves for the cities. Food crops often rotted in the fields in the absence of good roads or refrigeration technology. Planners faced a difficult choice between increasing investment in Siberia, which had oil, gas, coal, iron ore, and other resources but lacked the population and the infrastructure needed to develop them, and rebuilding the aged physical plant in the European part of the nation, where 70 percent of the population resided.[3] The arms race with the United States put added strains on the economy. Labor inputs for economic growth no longer sufficed to power the economy. Fertility was highest among the Moslem citizens of the Central Asian republics, while Slavic couples normally raised only one child. Leaders would not shift investment to Central Asia either.

But the already enfeebled Chernenko (who owed his career to Brezhnev, whom he had followed from postings in Moldova after the war into the Politburo) merely alluded to reforms and to the need for labor discipline, allegiance to the communist ideals of collectivism, and vigilance against the capitalist world. Upon the death of his mentor, he was kept out of power by anti-Brezhnev forces, who maneuvered to elect Yuri Andropov leader of the Party. But like Andropov, who ruled only briefly before succumbing to disease, Chernenko saw no need for significant change, although he advocated several changes in the way trade unions and higher education were administered. Reform, when there was any, came by administrative fiat, from the top down, and rarely with input from local managers, let alone workers or farmers. Indifference among workers indicated widespread dissatisfaction among the citizenry over the quality of life. Alcoholism remained a major problem. Though newspapers and official reports touted extensive scientific achievements, specialists with reformist leanings lamented bureaucratic barriers to innovation, international travel, and rapid publication, and the resulting declining performance.

The plodding gerontocracy was simply unable to consider such reforms as increasing the reliance on market mechanisms, or perhaps another New Economic Policy like that of the 1920s. They feared the devolution of initiative to factory managers, let alone to entrepreneurs (of whom there were few). Their worldview had been shaped by the Stalin years of fealty to the plans, struggles against internal and external enemies, and unshakeable emphasis on heavy industry and the military. Yet the fact that Gorbachev's appointment was announced the day after Chernenko's death suggests that the Politburo had already decided upon Gorbachev as the next leader when selecting the decrepit Chernenko. In December of 1984, addressing an ideological meeting of the Party in Moscow, Gorbachev already had referred to the importance of glasnost and perestroika. In the same month he had visited England, where Prime Minister Margaret Thatcher had said of him "This is someone with whom I can do business."[4]

Gorbachev's reformist attitudes toward the maladies of the Soviet system can be traced back to his days as a law student at Moscow State University, from which he graduated in the mid 1950s.[5] He and his closest friends, including his future wife, Raisa, came of age during the period of de-Stalinization. Many of the "children of the Twentieth Party Congress," including the sociologist Tatiana Zaslavskaia and the economist Abel Aganbegian, were also at Moscow State University in those days, as were the young biologists of a nascent environmental movement and other future reformers. Gorbachev joined the Komsomol at the university, and in the Komsomol and Party organizations from that time forward he was always the youngest person in any position to which he was appointed. He rejected many of Brezhnev's policies and approaches, although he did not speak openly about his disagreements, and he read *Novyi mir* (*New World*) and other liberal publications of the Soviet press.

Gorbachev set out to reform the USSR, including its science, after the virtual interregnum in policy making that characterized the last years of the Brezhnev regime. He set out to remove potential opposition to his reforms through what has been called a cold purge, booting scores of officials at every level of Party and government service (often 50 percent or more of the leading personnel) but never arresting or persecuting them. Simultaneously, he embarked on a series of reforms that, in some ways like those of the Khrushchev era, were often more impetuously raised than fully considered. Some observers argue that Gorbachev ultimately had to respond to increasingly contradictory pressures that he could not manage.

Gorbachev had to address persistent problems of scarcity, deficits, low-quality goods, and social inequities in society and economy. He also realized that the USSR could no longer and ought no longer to engage the United States in an expensive and futile arms race. He sought reforms in domestic and foreign policy in five interrelated ways. These reforms, as we will see, had direct significance for science and technology.

First, Gorbachev pursued perestroika. Perestroika required glasnost. Through open discussion of the failings of the Soviet system—the reasons for poor economic performance, rampant corruption among Party officials, the poor quality of goods and services, the failures of the Party during the Stalin era and beyond, and so on—Gorbachev hoped to encourage public support for reforms. One solution that was absent from discussion was dismantling the system. Gorbachev had no intention of allowing other political parties to compete with the Communist Party. He refused to change the Soviet constitution to remove the reference to the Party as the "leading organization" in the country. Yet the reforms would give local managers and officials greater authority to make decisions about resource allocation, inputs, and outputs, even if the broad outlines and specific changes originated from above. The planned economy would retain its privileged position, although factory managers were increasingly put on a system of self-financing (khozrachet) to negotiate favorable prices for goods and services, with profits available to them for wage funds to reward good workers. Whether the reforms would contribute to increases in the productivity of capital or labor and whether they would encourage innovation remained open questions.

Three other aspects of the reforms were just as crucial for science. One was "democratization" of Party and state institutions to bring like-minded, reformist individuals into positions of responsibility. Gorbachev's cold purge was a partial success. In the sciences, however, conservative scientific administrators held on to their positions; the Academy of Sciences leadership was wary of reforms, especially those that might threaten the power of its administrators or put at risk some of the Academy's rich holdings of real estate, sanitariums, and other wealth. Alferov, who had joined the Komsomol during World War II with patriotic fervor, and who in the middle of his career had become a member of the Communist Party, was seen as a reformer, although a number of his contemporaries found him much too enthusiastic about the Party's virtues and blind to its errors. In view of his parents' devotion to Bolshevism, this is not surprising.

Gorbachev also touted "new thinking" (novoe myshlenie), a rubric that invited new approaches and attitudes in foreign policy. The internationalist Alferov welcomed "new thinking." Rather than an arms race, with both the Soviet Union and the United States spending obscenely on nuclear and biological weapons, Gorbachev sought workable treaties with Ronald Reagan, eventually convincing the American president that he truly sought peace.[6] In the 1990s the US followed the USSR's lead in cessation of nuclear weapons tests, the Cold War ended, and Gorbachev earned a Nobel Peace Prize for his efforts. "New thinking" opened the USSR as never before to cultural, educational, and other exchange programs, and to joint ventures in the economic sphere. American and Soviet citizens enjoyed access to one another as never before.

The architects of perestroika thought in broad strokes of reform. They had no idea how specifically to make the economy, including science, a more responsive, innovative, and powerful tool to impel social progress. Bureaucrats, who had long treated science as a branch of the economy instead of a unique and dynamic social institution, hit upon various measures, regulations, and strictures to force scientific research to be more "economical" or effective. But rapid inflation and declines in production soon consumed the value of the ruble and triggered a deep crisis in science. Institutes lost the ability to pay salaries, to run experiments, and even to pay for heating. The results for science were devastating. For example, a number of valuable collections of plants at several biological gardens and institutes perished by the early 1990s, especially in research centers far from the concentration of resources in the major cities of Moscow, Leningrad, Kiev, and Novosibirsk.

Many research institutes hired economists to help them through the morass of reforms and rules inspired by perestroika. The LFTI hired a chief economist to assist the deputy director in coping with rapid change. And Alferov hired Iurii Kovalchuk (today a leading official of the Rossiia Bank) and Andrei Fursenko (now Minister of Science and Education) to suggest ways for the LFTI to take advantage of perestroika. Alferov focused his energies on ensuring budgetary allocations at the highest levels of government, using his contacts and his negotiating skills to secure funding for establishment of a microelectronics center at the LFTI. He worked closely with staff employees to understand the significance of economic reforms for his institute. Unfortunately, the political and economic crises that sprang from perestroika prevented any workable policy from being enacted.

The last aspect of the Gorbachev reform program was "speeding up" (uskorenie), which meant administrative and organizational measures to accelerate scientific advances into the economy. In some ways this harkened back to past efforts at economic reform by emphasizing the importance of innovation and by calling on managers, engineers, and scientists to do more to innovate, but giving them neither incentives nor appropriate organizational vehicles. Finally, glasnost meant discussion of several of the administrative weaknesses of Soviet science, many of which scientists had long tried to address, although not in any systematic way.

For scientists, uskorenie had important antecedents. From the late 1920s on, scientists and engineers faced mounting pressure to avoid "ivory tower reasoning" or "science for science's sake." They were supposed to be devoted to the goal of increasing output as rapidly as possible of benefit to the working class, and to predict concretely in one- and five-year plans of research what they would accomplish. Under Stalin, if they missed targets, they risked being accused of being inattentive and incompetent, and perhaps even of "wrecking." Most of them grew adept at demonstrating the tie between science and production. They reasonably claimed, for example, that in high-energy physics the design of powerful accelerators led to medical and agricultural applications. Work in nuclear physics begat cancer treatments, food sterilization, and reactors. Solid state physics led to myriad applications in energy, communications, and other technologies, although, as Alferov would point out, with a lag of a few decades when compared with the immediate applications in nuclear energy. By the late 1950s, this pressure had abated somewhat, and scientists had reestablished the level of autonomy needed to create new institutes, schools, programs, and "cities of science." Uskorenie signaled another effort to new harness science to improve economic performance. By January of 1987, however, Gorbachev had abandoned the slogan "uskorenie" and turned his focus to de-Stalinization.

There were at least two stages in the evolution of Gorbachev's basic economic approach. (There is a debate as to whether he planned it in two stages from the beginning or changed his strategies as he went along. The latter seems more likely.) Initially, he was Andropovian in his approach; that is, he was fairly conservative and emphasized the usual tactics of reform in words only, with an emphasis on uskorenie, the hope for rapid technological innovation to overcome the weaknesses of the economy. From about 1987 on, he took a more radical decentralized approach; in that year, the Party passed a law on state enterprises and legalized cooperatives and leasing, in part through self-financing. These two developments broke the central planning system and contributed to the decline of the Soviet economy. It is hard to judge precisely how these changes affected science, but according to one observer "cooperatives and leasing certainly created some havoc."[7] At the LFTI, as at many other institutes, a number of scientists rushed to turn their laboratories into profit-making units through businesses on the side. This often disturbed institute directors (who reasonably worried that scientists were stealing time and supplies from the institute) and committed communists (who maintained a strong distaste for anything that had the whiff of private enterprise).

Stephen Fortescue (a longtime specialist in Soviet science policy) has said that Gorbachev, in his early, more conservative "pro-technology" phase, expected that science would inevitably be useful to the economy. Then, as economic and budgetary chaos set in, the message from the government to scientists may have become "we don't have the money to support you, so do what you like, but presumably if you want to finance yourselves, you will have to be useful."[8] In any event, initially Gorbachev supported big science and technology, with the usual words about "speeding up" and "rapid assimilation" of results in production, and he backed such white elephants as the mega-project to transfer vast quantities of water from Siberian rivers into the Volga and Central Asian basins. The latter project fell under attack in such literary journals as *Novyi Mir* and was abandoned as costly and environmentally unsound.[9]

Ultimately, Gorbachev could not control the political and economic forces unleashed by perestroika. In elections to the Supreme Soviet that he called to involve the masses in perestroika, the people often voted against leading Communist officials even if they ran unopposed. In the first elections, Communist Party organizations were given the right to designate candidates for election as deputies to ensure a Communist majority in the Supreme Soviet. These seemingly loyal organizations were allocated one-third of the total number of deputies (750 of 2,250). But citizens sometimes rejected those candidates outright. When the Academy of Science slate excluded Andrei Sakharov, hundreds of mid-level scientists (especially physicists) from institutes in Moscow marched to the presidium building to protest. Sakharov was then added to the slate and easily elected. Alferov was also elected from the Academy slate of candidates; unlike Sakharov, he had been included in the original list as a mainstream communist. He continues his service in the Russian Parliament today.

Yet many intellectuals, who had been the bulwark of support for Gorbachev in the first years of perestroika, became dissatisfied with his unwillingness to permit opposition political parties to form, and rejected his heavy-handedness in running the daily deliberations of the Congress of People's Deputies. In August of 1991, hard-line communists attempted a coup. They placed Gorbachev (then on vacation in Crimea) under house arrest, declared a state of emergency, and ordered troops into Moscow. Boris Yeltsin and other officials rallied the people of Moscow in opposition, and the troops wavered. The coup failed, and its leaders were arrested. But on December 25, Gorbachev resigned as president of the USSR, and on December 31 the USSR disappeared. The former republics of the USSR, and the Baltic states (which the United States had never recognized as parts of the Soviet empire), became independent. Boris Yeltsin, Russia's president, now presided over many of the resources of the former USSR—including the vast majority of its scientific research institutes—and was a major figure in international affairs. Zhores Alferov and other Academy scientists were left scrambling to find financial and moral support to save the Soviet scientific legacy in these institutes.

Alferov and Science Policy in the Gorbachev Years

What did the Gorbachev reforms mean for science and for the career of Zhores Alferov? Glasnost—consideration of errors, mistakes, and poorly understood reasons for the poor performance—enabled public discussion of the strengths and weaknesses of the R&D apparatus. Scientists, intellectuals, journalists, and citizens considered anew the impact of ideological campaigns, top-heavy bureaucratic management, heavy-handed controls on travel, literature, and degrees, and overriding emphasis on military applications. From newly opened archives they learned in great detail about how Lysenkoism had handicapped the development of Soviet genetics until the 1960s, and about similar ideologically motivated attacks on cybernetics, chemistry, and physics.[10] At long last, the requirement that scientists acknowledge the role of Marxism-Leninism in their training, their research, and perhaps even their algorithms ceased. Marxist "study circles" disappeared, and Marxism itself lost resonance in the Higher Attestation

Commission (VAK). Some scientists openly questioned the predominant emphasis on applied science and the denigration of basic research. Many of the scientists who had considered ideological controversies to be a great inconvenience but a price of doing research in the USSR had simply avoided these discussions in earlier years; they also stayed out of the renewed debate about the costs of philosophical intrusion on science. So long as they received sufficient funding and were free to do research, they believed that was all that mattered. The approaching financial crisis caught their attention, as it did Alferov's.

Another prominent revolution in science during the Gorbachev years concerned a growing network of international contacts. Slowly at first, then in large numbers, scientists took advantage of new freedoms to travel abroad to participate in conferences, undertake joint research, and accept fellowships. The only limit seemed to be the willingness of Western colleagues, foundations, and funds to buy tickets and pay for accommodations. Autarky in science, whose roots dated to the rise of Stalin in the 1930s, ended. Physicists, including Alferov, rapidly expanded their contacts with European and American colleagues.

Yet travel to and contacts with the West, especially the United States, remained problematic, and in the view of many observers, including Alferov, bureaucratic obstacles to contacts contributed to the stultification of research. When the Soviet authorities permitted citizens to travel outside the USSR, they often did so at the last moment, although increasingly out of incompetence rather than ill will. Few scientists had the opportunity to accept invitations to foreign workshops, let alone make long visits such as Alferov's semester in Illinois. Invitations still passed through various postal inspectors and through the KGB agent in each institute. An invitation to a meeting often took months to reach a researcher, and often it arrived after the meeting was over. And scientific administrators often were sent to meetings instead of the invited scientists.

Vitalii Ginzburg documented the bureaucratic barriers to foreign travel in an essay published in a 1988 paean to perestroika, *Inovo Ne Dano* (meaning There Is No Other Way). Such leading proponents of reform in the USSR as the sociologist Tatiana Zaslavskaia, the physicist Andrei Sakharov, the *Novyi Mir* editor Sergei Zalygin, and the writer Danil Granin joined Ginzburg in defense of perestroika in that volume.[11] Ginzburg, like Alferov a member of the Communist Party, supported the Gorbachev reforms, although growing personal animosity between the two men prevented them from cooperating in pursuit of perestroika.

In his essay, Ginzburg sharply criticized the bureaucracies of science that were geared to isolating it, not serving it. He attacked the incompetence of the KGB and other officials who interfered with the smooth operation of R&D through their meddling. He singled out the "paper bureaucracy" that stifled science by requiring signatures for everything as a "malicious vile creature, a source of heart attacks, sorrow and tears, not to mention loss of time, ineffectiveness in work and so on." Ginzburg estimated that a million pieces of paper made their way through the chancellery of the Academy per year, including accounts, plans, declarations, residence permits, apartment permission slips, articles, defenses of dissertations, and records of foreign travel. In his institute, at least nine signatures were required before an article could be forwarded for publication. Of course, bureaucrats were worried that state secrets or information with great economic potential might find their way into foreign hands. But the system prevented all information from moving at other than glacial speed.

Having gotten documentation and permission forms from the presidium of the Academy to publish an article or read a paper abroad, the author then had to send copies to Glavlit, the main censorship organization.[12] Ginzburg ridiculed Soviet attitudes toward copying machines, which revolutionized research and writing in other countries. In the USSR one needed several signatures to copy articles, and the machines were centrally controlled, were open only a few hours per week. and could be operated only by special personnel.[13] On top of all this, many journals were printed in small runs, with limited page counts, and publishing a time-sensitive article might take eighteen months.

Even worse, one had to present and process 53 documents in order to travel abroad, and processing the entire folder would take several months. Ginzburg observed sarcastically that, after a brilliant effort on the part of the presidium of the Academy, the number of documents required was reduced to "only" 42. On top of this, for every trip a potential traveler had to fill out a ten-copy questionnaire that asked about his or her places of work and residence, marital status, previous spouses, children, siblings, and parents.[14] Of course, one could not make international phone calls or receive faxes without getting special permission and ordering the calls ahead of time. Finally, the permission granters seemed to engage in simple ridicule. In 1984, the Polish Physical Society awarded Ginzburg its Marian Smoluchowski Medal, but the president of the society informed Ginzburg that the award was being held up on the Soviet side, where Academy of Science officials were "considering" whether he might be allowed to accept it.[15]

In July of 1987, Alferov—who embraced perestroika for science—became the LFTI's fifth director in 70 years. Four months later, in a sense following Gorbachev's path of accelerating reforms, he conducted a cold purge of his institute. He replaced all scientific (as opposed to administrative) deputy directors, and instructed the new appointees to consider how to support science in rapidly changing economic and political conditions, including how to determine the implications of khozraschet (the requirement for self-financing) for scientific research. As inflation started to gobble up government financing, scientists found themselves forced to turn, for funding, to contracts, joint projects with foreign colleagues, and a limited supply of grants.

Scientists had rarely considered the economics of science. They had gotten most of their funding through relatively stable line-item budgets, not competitive grants. Soviet economists, for their part, had a mercenary and mechanical understanding of science. The result was that institute after institute, in order to be certain of annual funding, had to demonstrate, in new ways with new forms, that its research was economically important. Institutes had become "bookkeepers drowning in a sea of paper." As resources became tighter, fewer funds were available for research. Supernumerary employees became a drag on the overall salary fund. The volume of applications for project financing grew more rapidly than the staff available to complete the paperwork. As inflation ate up finances, Andrei Fursenko, the LFTI's deputy director for science, admitted he had no choice but to freeze all hiring and all accounts. Each week he had to decline some request, at the cost of increasing the unpopularity of the administration. To deal with economic uncertainty, the LFTI hired first one economist then another to create a modern comptroller's office with a staff capable of meeting new and constantly changing rules set for academy and branch research institutes by the Council of Ministers. These rules failed to include anything like an overhead rate for government contracts. At the LFTI, the administration set out to determine what percentage of the institute's budget went to the salary fund. It was only 18 percent, versus 28 percent for comparable institutes. Without resolving the question of how to raise that rate for contracts, the administration eventually decided to cut staff, which would permit them to increase salaries among qualified staff. But some departments resisted; they refused to fire people to increase wages.[16]

Scientists experimented with a new form of research organization: the cooperative, a kind of alternative to stodgy academic laboratories. Cooperatives organized in the late 1980s to take advantage of an atmosphere of uncertainty about what was permissible and what was not. Scientists in a variety of fields sought to gain funding and autonomy in these co-operatives. Alferov was rightly skeptical toward the cooperatives, very few of which succeeded: "If these scientific cooperatives are created as fully independent organizations which offer new scientific products and realize new ideas which the current academic structure slows down—thank heavens!" Unfortunately, most of the cooperatives used institutes' laboratory space, apparatus, materials, and even staff without giving anything in return, let alone paying rent.[17]

Financial challenges tempered the excitement of perestroika in research institutes. Eduard Tropp, scientific secretary of the presidium of the St. Petersburg Scientific Center since 1994, has referred to 1989 as "a great time, a time of hopes and a time of optimism." Yet many scientists had a feeling of impending crisis. The hope was connected with the belief that reform of the Communist Party would lead to a better understanding among bureaucrats of the importance of support of basic science, the growing international contacts among scholars, and the dissipating threat of nuclear war. Scientists of the Leningrad Scientific Center indicated their hope in a variety of ways. Whereas the Center's annual report for 1989 listed successful fulfillment of the annual plan according to directives of the Communist Party, its 1990 report did not mention the Party, but rather referred to achievements "in raising the effectiveness of center workers and research of the institutions of the center in conditions of perestroika and democratization of Soviet society."[18]

Politically, perestroika promised openness and restructuring; however, Alferov and other institute directors discovered that it created daunting economic problems that slowed scientific activity. During the Brezhnev era, as the Soviet economy slowed and revealed far too many obstacles to innovation, government officials considered science largely for its economic and military potential. Paradoxically, in the environment of perestroika, officials came to see science even more exclusively as an economic lever, and questioned the value of basic research—science for the sake of science. This meant that laboratory and institute directors now had to watch expenses closely. At the LFTI and at other leading institutes, they had at one time expected requests for more staff, new facilities, and modern equipment to be met as a matter of course. Tightened budgets required both an unpleasant reckoning with redundant personnel and a burdensome search for new sources of income—for example, contracts with other organizations, including the military. In some cases, institute directors were forced to curtail research programs. They could not fill vacancies, although many refused to fire redundant personnel for a variety of reasons. They often knew many of the staff members personally, even in an institute as large as the LFTI. They felt loyalty to senior scholars whose days as productive scientists might have passed but who had been close associates for decades. The entire Soviet labor system was based on the concept of full employment. Even where overemployment resulted in the allocation of a task that one person could accomplish comfortably to two or more individuals, employers were psychologically loath to cut costs by letting staff members go.[19]

Economic uncertainties generated new pressure to demonstrate prematurely the tie between fundamental research and applications. But, as the case of the LFTI indicates, scientists faced the continued bind of having to serve as researchers while being seen as the vehicles for introducing innovations in industry, and even such modern industries as microelectronics had begun to slow and contract along with the rest of the economy. Tuchkevich, Alferov, and others had long been disturbed by the inability of Soviet and now Russian industry to produce semiconductors in the quantity and quality that other nations managed to produce. In November of 1986, Tuchkevich wrote to the Academy's president, Gurii Marchuk, to complain about the "unsatisfactory tempos of development of power semiconductor instrument building in Minelektroprom [the Ministry of the Electronics Industry]." In the eleventh five-year plan, production had dropped more than twofold in comparison with the ninth and tenth five-year plans. Tuchkevich observed that the lag in production would worsen as demand increased, and that the problem involved not only the numbers of devices manufactured but also "the type, parameters, level of technology and . . . quality" in comparison with devices produced abroad. The devices in question included thyristors, diodes, integrated circuits, and instruments. Such devices were crucial for safe and inexpensive operation of municipal and railway transport, for computer and information technology, and for controlling nuclear and metallurgical processes. As an afterthought, Tuchkevich mentioned the lag in introducing these devices in the consumer goods and agricultural sectors. Apparently he did not recognize the importance of demand from those sectors in stimulating production. Instead, microelectronics production and distribution relied on planners' preferences and decrees. Tuchkevich called for the creation of several national centers to produce devices under the direction of the LFTI and other leading physics centers.[20] But what could the LFTI have done about the production of semiconductor devices? Wasn't that the responsibility of industry? And whose fault was poor production?

Party officials thought they had the answers. Much of the effort to reform science in the 1980s revolved around a new series of campaigns.

Grigorii Romanov, the Leningrad Party leader, brought full pressure to bear on the Academy to reorient the research of Academy institutes toward the needs of local industrial production by establishing a Leningrad Scientific Center. He and other officials sought to foster technological innovation, although largely by administrative fiat rather than through increased resources of finance or manpower. The Party's regional first secretary, Lev Zaikov, launched Intensification-90, a campaign that Gorbachev endorsed upon becoming general secretary.[21] Along with Intensification-90, scientists gained research funding through such regional scientific-economic programs as Ladoga, Science-90, Black Soil, and Health.

Intensification-90 produced a flood of proposals to accelerate the achievements of science and technology into the economy. On October 28, 1986, Tuchkevich wrote the general director of the Svetlana factory, O. V. Filatov, with a proposal to organize a "temporary" three-to-five-year effort to produce powerful new semiconductor devices. The laboratory of I. V. Grekhov had developed new types of thyristors and transistors and instruments based on them. Tuchkevich suggested work on powerful high-frequency electronics with a frequency range from scores of kilohertz to scores of megahertz and applications in radio communications, plasmochemistry, and so on.[22] Yet Intensification-90 seems ultimately to have generated only paper, not innovation, with ministerial barriers again preventing innovation and with institutes again being pushed to become ambassadors of science in industry.

Several reforms were successful: increasing openness in scientific publication, easier access to foreign travel, some democratization in the running of research institutes. But in general, Party officials in the R&D apparatus were unwilling or financially unable to permit real autonomy for research institutes, while most laboratory and institute directors saw as the solution to any problem more money and more staff, never any kind of administrative reform.

Streamlining the procedure for Candidate of Science and Doctor of Science defenses might have been considered; it might have indicated greater trust in scientists at the institute level. The Higher Attestation Commission would have lost power, and local bodies of scientists would have gained greater autonomy. Instead, defenses lagged under the weight of such bureaucratic impediments as the dozens of signatures and copies of each document needed and a system that sought to verify the political reliability of the scholars in question. By the end of the Soviet period, only one of three people in the Russian Federation who finished graduate research defended a Candidate of Science degree, and less than one-third of young specialists were able to find employment in their area of specialization.

When Alferov became chairman of the Leningrad Scientific Center, in 1989, he determined to use his authority to pursue a variety of reforms in research and in higher education, and to steer the Center and its Academy institutes back toward basic research. Alferov hoped the Gorbachev reforms would lead to a less formal and more dynamic system of higher education, one better equipped to train young scholars for careers in science. He embraced perestroika and glasnost for their potential to rejuvenate basic research in the USSR. In a series of speeches in the Parliament and at academic institutions, and in articles in newspapers and journals, he promoted reform. Sensitive to the pressure on scientists to contribute to the economy, especially since Gorbachev's call for uskorenie, he argued for great care before adopting rules requiring economic accountability among researchers in basic science. He worried about the need to restore the prestige of science, which had declined during the Brezhnev era. The significance of perestroika, he suggested, lay precisely in recognizing the value of science for its own sake. He saw science as the true measure of society. "Fundamental science," he wrote, "establishes that level . . . of society and civilization as a whole toward which we should strive."[23]

Alferov believed that perestroika would lead to the rejuvenation of a variety of fields, from nuclear energy and microelectronics to more theoretical pursuits such as his own field of research. He pointed to successes in these areas as evidence that the government should at least support basic research morally and psychologically, if not with significantly increased funding. A narrow, pragmatic, economic view of science would slow development. Alferov called for patience, not pressure for economic results, recognizing that basic research might lead to applications only decades after a discovery.[24] He hoped that perestroika might ease some of the reporting and planning requirements that interfered with the conduct of science. He complained that planning mechanisms had become outdated. There were too many plans—quarterly, annual, five-year—when the latter alone would suffice. And he called for the specialists alone, not government bureaucrats, to set directions for research.[25]

Disjunctions and bureaucratic barriers between education and research troubled Alferov. They would not be overcome during the Gorbachev era. Alferov attacked the "formalism" and "bureaucratism" that interfered with the formation of a symbiotic relationship. This problem started, for an individual, in high school, and persisted through the completion of an advanced degree. Alferov believed that the requirement to defend both a

Candidate of Science thesis and a Doctor of Science thesis was needless. He acknowledged that the second thesis requirement may have been useful to raise the quality of cadres in the 1920s and the 1930s, but by the 1950s it had become a waste of time. Rarely, he asserted, did anyone even remember the theme of a doctoral dissertation, and such dissertations often were written by gathering and reworking a series of published papers. Alferov did not address the fact that the second Soviet dissertation involved another check on the political qualifications of candidates, since the VAK might reject a doctoral committee's recommendation. In the view of many people, the VAK simply served as the final gatekeeper of Soviet orthodoxy.

Alferov worried about the genuflection before authority that was connected with "formalism." He acknowledged that it existed as much at the LFTI as at other institutes. More than scholars in other countries, Soviet scholars found themselves scrambling to cite big names in their fields to advance their careers. Soviet mass publications and scientific journals contributed to this through hero worship. Many of the big names of science had hundreds of publications, having gained the privilege of attaching their names to articles in whose writing (and research) they may have had little involvement. Alferov claimed this problem was not as extensive at the LFTI, which was a more democratic institute. "The aspiration to gain approval of the academician, the academic secretary and so on—from my point of view," Alferov wrote, "it's necessary to recognize this weakness."[26]

Alferov seems to have believed that increasing funding would go a long way toward solving most the problems facing basic research. At the nineteenth Party conference, in June of 1988, Gorbachev was distracted by growing opposition in the Party to his reforms and by efforts to prepare for elections to the Congress of People's Deputies. But Gorbachev mentioned science in his keynote address, and he released financial data indicating that fundamental science was at a critical juncture. Alferov echoed this view: "The situation worries all of us who are involved in science. I was convinced of that at a meeting in our institute on the eve of the conference. There were so many observations, desires and suggestions!" Institute physicists suggested dozens of ways to make up for the financial shortfalls. They fretted about a significant lag in scientific instrument building that required them to purchase devices abroad. This put a great drain on their hard-currency reserves. But the need to conduct research in such areas of importance to the institute as quantum electronics and semiconductor physics left no alternative. Alferov worried that financial and equipment shortfalls in laboratories of second-tier institutes and those in the union republics were having a disproportionately greater impact on provincial science.[27]

Alferov as Institute Director and Scientific Administrator

Every administrator knows that joys and sorrows accompany the position. An institute director may pursue a vision of research, organize the scientific collective in pursuit of that aim, and ensure the stability of institute personnel and programs. The director gains recognition in the field, and his salary may reward him for his efforts. Most directors simultaneously complain about their responsibilities. They often interfere with the director's own research program. He must focus on fundraising and on ensuring good relations with government officials and funding agencies. And he may need great wisdom to adjudicate personnel problems.

In the Brezhnev era, the position of institute director required more and more time to raise funding for institute coffers and less time to undertake research. Yet as important as political connections were to the success of a director, his staff, and the institute's research program, an institute director could not have reached that positions without a career of scientific excellence and administrative verve. Political reliability—allegiance to the Communist Party's dictates and programs—was less important, though about 70 percent of institute directors (Alferov included) were members of the Party. According to a number of critics, several institute directors relied too heavily on connections and authority within the Party in shaping the direction of institute policies, and on using those connections and that authority to dominate a field of research. They singled out Nikolai Basov of the Physics Institute of the Academy of Sciences for his single-minded devotion to laser research, contending that he attempted to shape all institute programs in support of laser R&D, to the detriment of other programs.

Through the early 1980s, Alferov and other academy institute directors could be almost certain that their requests for increased funding or for more research personnel would be approved. Billions of rubles were available for research in the search for technological breakthroughs in the race for parity with the United States in the Cold War. The LFTI got its share of rubles for research on missile defense, fusion, lasers, and communications. That was the heyday of Soviet science. Money seemed to grow on trees or in snow banks, at least in comparison with the last two years of Soviet power and the 1990s, when real budgets shrank by factors of 20–30. Yet physicists had to turn increasingly to sources of funding outside of budget line items, often in concert with branch industry institutes, to expand research programs. To this day, older physicists nostalgically remember the joys of stable line-item funding, which left them without the need to apply for grants continually as American researchers do.

Alferov was successful in the effort to generate funding through contracts. For example, in 1988, in response to the approval of a list of projects published by the Central Committee of the Communist Party connected with the development of new spacecraft, Alferov wrote to the deputy director of the scientific-organizational department of the Academy of Sciences and the general director of the so-called scientific-production organization Tochnost' (meaning Precision) about the LFTI's determination to participate in a project to develop highly sensitive radiation-stable multi-spectral matrix photoreceivers. The LFTI would join a firm called Elektron and the Leningrad Polytechnical Institute in the effort, with the understanding that the funding required was on the order of 9.5 million rubles and the term of work was roughly three years. (Elektron itself dated to the 1930s as a research institute of radioelectronics, was evacuated during the war, and was reopened in 1946. In 1971 it became a scientific production organization. It survives as TsNII Elektron.) The contract also required 26 additional staffers, including young scholars, to be added in the next four years.

When it came to expansion of research programs for basic research, institute directors worked closely with Academy leadership. Alferov turned to A. M. Prokhorov (a Nobel laureate in 1964 for his work with Nikolai Basov on the laser, now academic secretary of the Academy's division of general physics and astronomy) for approval to establish a temporary laboratory of high-temperature silicon carbide electronics. Under V. E. Chelnokov, the institute had successfully conducted research on developing these devices, but technical problems interfered with further results. The production of silicon carbide required working temperatures of 500–800°C, and the creation of semiconductor devices at those temperatures on the basis of silicon and A^3B^5 compounds was in principle impossible. The goal of the temporary laboratory was intensification of research on the manufacture of the devices and their rapid introduction into industry. Alferov requested 200,000 rubles per year for three years from the Ministry of General Machine Building, whose agreement he apparently had secured.[28]

Work on controlled thermonuclear synthesis (fusion) has been prestigious among Academy scientists since the early 1950s, when Igor Tamm and Andrei Sakharov first advanced the theoretical design for the tokamak reactor. LFTI researchers had contributed significantly to this field, too. Research took off after Igor Kurchatov succeeded in convincing Nikita Khrushchev to declassify the program to allow it to be discussed at the first Geneva Conference on the Peaceful Uses of Atomic Energy, in 1955. The fusion program was centered at the Kurchatov Institute in Moscow, and its major champion in the last years of Soviet power was Academician Evgenii

Velikhov, head of that institute and one of the Academy's vice presidents. Velikhov has ensured continued Russian participation in the International Fusion Research Program (ITER), from the uncertain days of minuscule budgets in the early and mid 1990s to the rejuvenated program that will see a reactor built by 2015. Other institutes, including the LFTI, the Ukrainian Physical Technical Institute in Kharkiv, and the Institute of Nuclear Physics in Akademgorodok, also conducted fusion research, often on alternatives to the tokamak—the probkotron, open traps, and so on.

Working closely with such specialists in fusion research as Viktor Golant at the LFTI, Alferov sought as director to build on the national enthusiasm for fusion research to expand his institute's programs. The LFTI had earlier received a huge award to erect a fusion complex for the Tuman tokamak that took up an entire building and a quarter of a block across the street from the Politekhnicheskaia subway station. In July of 1988, Alferov wrote to the deputy president of the Academy of Sciences for capital construction, enclosing a letter he had sent to Academy president Gurii Marchuk endorsing the building of a new tokamak called the IFT. The Leningrad regional energy company had approved the technical specifications of the project, and the main city planner, S. I. Sokolov, had approved a new building with a floor plan of 2,000 square meters next to the existing Tuman building. Alferov asked that the Academy authorize two to three million rubles for the construction work and standard equipment, construction to be undertaken in 1991 and 1992.[29] Unfortunately, at precisely this time the Soviet Union broke apart. In the first years of economic turmoil under Russian President Boris Yeltsin, funding for this project dried up.

Alferov acknowledged the increasing programmatic and psychological challenges of being an institute director when facing this constant search for support for research in an atmosphere of growing political uncertainty. The time-consuming effort prevented him from engaging in his own work. He constantly thought about his ideas on semiconductor heterostructures, frequently discussing them with staff members of his lab and especially enjoyed pondering them on his days at his home on the Gulf of Finland at Komarovo. He grew excited about superconductivity and then nanostructures. His musings led him to consider how long he would remain an institute director. Being a director had become physically draining, too, and he had yet to give up cigarettes and long nights.

Alferov, a well-rounded man with interests in the arts and literature, turned always to his favorite authors to seek respite from the pressures of administering an institute and a center and the stresses of parliamentary responsibilities. He continued his exploration of the history of World War II, and he read and re-read *Life and Fate*, Vasilii Grossman's 1960 novel about a physicist and his family before, during, and after the battle of Stalingrad. Grossman, a news correspondent for the Red Army's newspaper *Krasnaia zvezda*, spent the war on the front lines, filing stories on the great battles and on the lives of ordinary soldiers.[30] During the war, mothers, fathers, wives, sisters, brothers. and children of soldiers read his stories eagerly. But *Life and Fate* saw light only many years later. Because it touched on topics the Soviet leadership considered sensitive, including the arbitrary decisions of unfeeling bureaucrats, anti-Semitism, and Stalinism's resemblance to fascism, KGB agents confiscated the manuscript, and in 1962 the Politburo official responsible for ideology, Mikhail Suslov, informed Grossman that he would never see his novel published, for it was even more dangerous than Boris Pasternak's *Doctor Zhivago*. Grossman died in 1964. His book was published in the West in 1980, and in 1988 it appeared in the USSR in serial form in the monthly journal *Oktiabr*, where Alferov and thousands of others read it. In this period of glasnost, Alferov and millions of other Soviet readers saw the publication of previously taboo books, essays, and poems, and not only in leading journals—in a weekly such as *Circus* or *Chess* one might find Mandel'shtam's "Ode To Stalin" or something on Trotsky, who had been written out of history during the Stalin era.

In April of 1989, Alferov gained another administrative position that he hoped to use to improve the state of research and development: chairman of the Leningrad Scientific Center (LNTs, now the St. Petersburg Scientific Center), a post he still holds at this writing. The LNTs comprises 70 institutions, organizations, enterprises, and scientific societies in the Leningrad region, 33 of them independent scientific research institutes, with a total of 13,200 employees. Several of the largest academy institutes in Russia fall within LNTs: the LFTI and the Konstantinov Nuclear Physics Institute, each with 2,200 staff members.[31] Chairing the LNTs has consumed a great deal of Alferov's attention.[32] When he was appointed to the position, Alferov claimed that his previous and consistent comments about dislike of administrative work and the superfluousness of the LNTs had been taken out of context. He had recently turned down the chairmanship of a national committee on vacuum technology. But he willingly accepted the position at LNTs because perestroika provided the perfect atmosphere for someone dedicated to reform of science, the protection of basic research, and the advancement of Leningrad's position as a leading scientific center.

Upon assuming the chairmanship, Alferov spoke publicly about his intentions to reform the LNTs. But significant reforms did not follow. There were two reasons. First, Alferov had been chairman for only two years when the Soviet Union ceased to exist. Second, Russia was immediately buffeted by the financial problems mentioned earlier in this chapter. Rather than reform, Alferov and his colleagues had to embrace a simpler goal, namely to save fundamental science. They lobbied the municipal, provincial, and federal governments incessantly, with limited results. But in the general atmosphere of consideration of the failings of the Soviet past, specialists had begun to question the Soviet development model and its contribution to severe environmental and other problems. The scientists of the LNTs recognized the need to study not only natural scientific problems but also societal concerns, using the tools of the social sciences and the humanities. They considered economic, ecological, and even ethical problems, with the participation of social scientists. During the Gorbachev era, the LNTs attempted to foster interdisciplinary research, training, and international programs through a special coordinating council with seven major research areas, ranging from the physical sciences to computing to ecology and the humanities.

Having reached the pinnacle of the Soviet scientific establishment, and facing a constantly changing political, economic, and scientific world, Alferov hoped to use his authority to defend science from pressures to focus excessively on applications and to secure funding across the entire range of disciplines represented in LNTs institutes. He hoped to abandon what came to be called "command-administrative" measures used to direct science in order to "return science to Leningrad and Leningrad to science." "Command-administrative" became the standard Gorbachev-era epithet, applied to Brezhnev-era programs. For example, Igor Alekseievich Glebov, the first chairman of the LNTs from 1983 to 1989—a specialist in electrical power stations, a graduate or the Leningrad Industrial Institute in 1938 and head of the Leningrad All-Union Scientific Research Institute of Electromachine Building—represented the Brezhnev-era pressures to conduct applied research and efforts to "perfect" the research enterprise. Perfection would arise not through reform but modest modifications of existing funding and administrative measures from above.[33] Still, many individuals remember Glebov as a good leader. He willingly followed the directives of the Leningrad Party Secretary, Grigorii Romanov, yet cared for science. He strove to protect the humanities, libraries, and archives, not only to serve big science, and he knew and remembered almost everyone's name.

Alferov said he stood for autonomous science, science that was based on its spiritual and social status in society, not science directed from above, not "command-administrative science." Through his appointment as chairman of LNTs, Alferov hoped to combat the "formalism [that] had penetrated into the Academy of Sciences." He wished to fight the centralization of decision making under inflexible plans that had slowed scientific progress. Although himself a member of the establishment, he believed that older scientists in Moscow tended to occupy the major positions in the administrative apparatus for Soviet science. He criticized the continued obstacles of sending young specialists abroad to study and do research. Of course, Alferov noted, the threat to science was in no way comparable to the challenges Ioffe faced in 1918 when he had to organize an institute, acquire foreign journals and purchase new equipment out of nothing during the Revolution and the civil war, nor with the period of the Stalinist purges.[34] Alferov—and seemingly all Leningrad specialists—fought the effort of the Party apparatus to turn the scientific center into a division of the Academy of Sciences, like the Urals division or some of the branches of the Siberian division, to meet strictly economic needs. In this attitude, Alferov and the others emphasized the importance of renewed financial and moral support for basic research.

Yet it must be pointed out that Alferov's efforts to paint himself as somehow representing reform obscures the fact that he was a member of the establishment. He was a leading member of the national scientific hierarchy, and the director of one of the largest institutes in the USSR. He belonged to a number of the Academy councils in Moscow that coordinated long-range research programs. Granted, the Moscow scientific establishment dominated these councils, and Moscow's Academy institutes had a privileged position, as did Moscow itself in terms of power and resources. In a word, significant regional tensions prevailed in Soviet science, with the Academy's Siberian division and its Urals center, and the republican academies of science, not to mention the Leningrad center, and Moscow institutes in which Alferov, as a proud Leningrader yet Academy leader, was often forced to play a conflicted role.

As a Leningrader, as a spokesman for that scientific community, Alferov referred to the great scientific heritage of his city as the cradle of Russian science centuries earlier. The administrative offices of the LNTs are in the Main Building of the Academy of Sciences, a magnificent nineteenth-century structure on the Neva River that served as the presidium building for the Academy itself until Stalin forced the scientists to move to Moscow in 1934. With its massive columns, it remains a monument to Peter the Great's efforts to establish the Imperial Academy of Russia in 1725. Peter modeled his Academy on the Royal Society in London and the French Academy. Alferov and other LNTs personnel take pride in the fact that the Center includes several institutes that date to the time of Peter: the Library of the Academy, the Botanical Institute, and the Kunstkamera, one of the world's first natural history museums.[35]

Peter the Great, and later such other Westernizers as the Bolsheviks, saw science as crucial to raising the level of Russia's culture. Peter was forced to import European scientists to fill his Academy in the absence of an indigenous scientific tradition and scientists, thus giving impetus to the concentration of honorific and research functions within the Russian Academy. The European academies largely kept honorific functions separate from independent research settings. The concentration of honorific positions and research in the Russian Academy of Sciences persists into the twenty-first century, and serves as another tension between Alferov the reformer and Alferov the member of the establishment.

Initially, Alferov said he intended to remain as the chairman of the LNTs for no more than five years, during which time he would try to revive its administration, he said, with scientists rather than bureaucrats directing science. Alferov criticized the "cult of directors" that had evolved in the LNTs, which indicated the tendency of policy for science in the Brezhnev era to follow a pattern of proclamations, edicts, and signatures issued from above. He brought new people into the presidium of the LNTs, all of them active laboratory directors doing world-class science. Revealing his personal enjoyment of the arts and his recognition of the importance of supporting activities in this area as well, he sought to raise the visibility of the humanities and social sciences that served "as a moral foundation for other branches." In February of 1988, a fire ravaged the library of the Academy of Sciences, damaging rare manuscripts. Alferov believed that a fire in a library was immeasurably worse than one in a laboratory, since irreplaceable national treasures had been destroyed, and that the failure to protect the library was an strong indication of the misplaced priorities of conservative bureaucrats who valued hard science above all else.

Another major way Alferov hoped to reform science was by fostering a new relationship between scientific and higher educational institutions to ensure the flow of well-trained young scholars into cutting-edge research. He proposed, for example, that Leningrad State University be "returned to bosom of the Academy of Sciences." He recalled that the Academy, as Peter the Great envisaged it, was a unitary organism with research at its core— the Academy itself—invigorated by a gymnasium (that is, a special high school). While Leningrad University would remain within the Ministry of Higher Education, Alferov was able to pursue his interest in a "contemporary synthesis of science and education" in other ways. He succeeded in establishing a lyceum that combined a heavy scientific curriculum with a liberal arts curriculum under the LFTI umbrella in 1988 to train promising students of junior high and high school age. LFTI physicists also created new or strengthened existing affiliated departments in Leningrad State University, the Polytechnical Institute, and the Leningrad Electrotechnical Institute. And in 2002, the St. Petersburg Academic Physical Technical University—that very institution inspired by Peter the Great—gained a charter to award physics and biology masters and PhD degrees. A number of these institutions became the core of Alferov's controversial Scientific Educational Center established after the award of his Nobel Prize.

Alferov, Sakharov, and the Congress of People's Deputies

Alferov gained another visible role in the political establishment in 1989. In addition to his administrative and scientific lives, he was elected as a deputy to the Congress of People's Deputies, the forerunner of today's Russian Duma (parliament). Perhaps 30 or 40 other scientists were also elected. But the activities of the Congress centered more on the big issues of perestroika and glasnost, the war in Afghanistan, and the appropriate future role of the Communist Party than on issues of science and technology directly, and the scientists never developed a lobby or a pressure group.

In the Congress of People's Deputies, Alferov came to play the role of protector of the Soviet legacy of science and technology. The fact that the deputies focused almost entirely on determining what perestroika meant and little on specific laws limited any chance he or others would have of improving the state of the scientific enterprise. In all, Alferov delivered only two major speeches in the Congress. The general theme of his speeches was the need to protect the great heritage of Soviet science in the face of political uncertainties and budgetary shortfalls. Alferov argued that it was possible to separate scientific research from the bureaucracy of officials who were supposed to administer it fairly but who in fact interfered with its smooth operation and denied it adequate support. Even in the atmosphere of raging debates over the future of perestroika, Alferov enunciated the belief that science was somehow an apolitical endeavor, and that the state ought to fund it at the appropriate level and give scientists complete autonomy to determine the direction of research. Many scientists have naively shared the view that it was possible to separate science from politics—for example, Germany specialists during World War I and during the rise of the Nazis, and, in the United States, nuclear physicists who sought to return quietly to their universities after the Manhattan Project.[36] Events in the Congress of People's Deputies would indicate the impossibility of ignoring the intensely national and personal politics of perestroika.

Alferov in the Russian Parliament, admonishing fellow deputies and the Putin administration for inadequate support for basic science, October 11, 2000, one day
after the announcement of his Nobel Prize.

Alferov entered the Congress at the same time as the physicist Andrei
Sakharov, the father of the Soviet hydrogen bomb, later a scientific activist and then a dissident, and finally the conscience of perestroika. Sakharov's background, career path, and determination to engage the limits of
perestroika—even to entertain the end of the USSR as perhaps a desirable
outcome—stand in contrast with Alferov's, and they help us to understand
how Alferov was a reformer only within the limits of the Soviet system. To
put it simply, Sakharov came from a privileged family and Alferov did not.
And yet both men were products of the Stalin era. For Alferov, the Soviet
system facilitated his parents' upward mobility and his rise to the top of the
scientific establishment, and the creation of the world's largest scientific
establishment was confirmation of the essential goodness of the Soviet system. Sakharov's first major steps in physics were in the military enterprise,
whose leaders and purposes Sakharov eventually rejected.

Sakharov's contribution to the Gorbachev era as the conscience of perestroika at a time when its possibilities briefly seemed limitless remind us
that Soviet physicists rarely engaged in civil society as political activists.
Few of them had the courage that enabled Sakharov to challenge political authority. Yet because of the fact that several leading scientists have

become dissidents in their societies, we often think that scientists are more likely than other individuals to be attracted to dissent because of the nature of their enterprise: they engage in the disinterested, skeptical investigation of natural phenomena in search of the truth. (Other examples include Galileo and Fang Lizhi.[37]) When confronted with institutions or polities that interfere with their autonomous pursuit of knowledge, or that mistreat large number of fellow citizens, they protest. In the United States, no sooner had they contributed to the development of nuclear weapons than many scientists joined national associations to work against the arms race—the Federation of American Scientists, the Union of Concerned Scientists, and others.

In the USSR, to be sure, a number of scientists, particularly physicists and mathematicians, joined the dissident movement in the Brezhnev era. Yet the numbers of these individuals were small, for scientists are also a product of the society in which they work. They are trained as members of an existing establishment, they are dedicated to their work, and many of them find extracurricular activities uninteresting, unnecessary, or even threatening to the scientific enterprise. They prefer to get regular funding, do their work, and avoid politics. Alferov avoided the politics of dissidence, had few colleagues or friends who actively engaged in dissent, and did not himself contribute to the movement. Although he earned a reputation as a member of the scientific and Party establishment in the 1960s and the 1970s, if not a stickler for orthodoxy, he disliked confrontation and heavy-handedness. Finally, he saw in the examples of Konstantinov and Tuchkevich the need for circumspection. As institute director, he did not punish individuals whom the authorities identified as dissidents or troublemakers. So although he was a liberal by the standards of high politics of the USSR, and in his desire for reform of the scientific enterprise, he had no interest in seeing the Communist Party lose its leading role or in abandoning the centrally planned economy for a market system. Not so Andrei Sakharov, whose example as the conscience of dissent and perestroika stands out in the grayness of Soviet politics, and whose activities in the Gorbachev era contrast sharply with those of Alferov—and the vast majority of scientists.

On a dreary day in December of 1989, thousands of citizens in scores of cities across the former Soviet Union gathered to march in Sakharov's honor. Many of these people believed that the era of political reform, of perestroika and glasnost under Mikhail Gorbachev, had evaporated when Sakharov died. Alferov attended Sakharov's funeral, aware that a great man had left the Russian political arena, but still believing that perestroika would run its successful course, and not disturbed that Gorbachev had decided not to postpone the work of the Congress. Instead the deputies had observed a moment of silence.

Andrei Sakharov was an enigma: a product of the Stalin era, yet a political liberal; a family man, yet distant to his children and colleagues; the developer of the most powerful nuclear weapons ever built, but devoted to arms control.[38] He entered the public arena in the 1960s as an uncompromising human rights activist. His constant criticisms of the Brezhnev regime landed him in internal exile for six years. Gorbachev finally invited him to return to his home in Moscow in January of 1986, and to resume work, Gorbachev said, for the sake of the motherland.

Alferov's parents despised the reactionary tsarist regime; Sakharov's parents, members of the liberal intelligentsia, were happy to see Nicholas abdicate, but not to see the tsar and his family murdered by the Bolsheviks. Sakharov's father, Dmitrii, studied medicine and then physics at Moscow University, where Andrei eventually studied. His mother was the daughter of a general and a member of the nobility. Sakharov was born in May of 1921. From his youth he was a loner. He had few close friends, preferring to read a book rather than to go out. He was aware of great changes that occurred under Stalin, of glorious industrialization and collectivization, and of the famine in Ukraine in 1932–33, and arbitrary arrests including that of his uncle. He focused on his studies, earning honors and the admiration of his fellow students. He entered Moscow State University (MGU) in 1938 at the age of 17, perhaps attracted by rapid developments in theoretical and nuclear physics. At the university Sakharov no doubt noted the philosophical and political intrigues that plagued physics, but he stuck to his studies and avoided controversies.

After the German invasion of June 21, 1941, the authorities evacuated scientists, students, and equipment from Moscow, Leningrad, and other cities. Sakharov was transferred to a radio factory in Ashkabad, Turkmenistan, then to a munitions factory in Ulyanovsk (Simbirsk), where he met his first wife, Klavdiia. In December of 1944 he received an invitation to join the Moscow-based Physics Institute of the Academy of Sciences (FIAN) graduate program. Upon arrival in Moscow, he went straight to Igor Tamm, a Nobel laureate in 1958 (along with Pavel Cherenkov and Ilya Frank) for his work on the Cherenkov electromagnetic effect. Sakharov worked primarily on elementary particle physics. Suddenly, he and hundreds of other promising young scientists were drafted into the nuclear bomb effort. In June of 1948, Tamm told Sakharov that he, Sakharov, and several others had been ordered to work on a hydrogen bomb. The changes at FIAN, like those at the LFTI, were immediate. The relaxed atmosphere to which physicists were accustomed suddenly gave way to strict controls, and the NKVD increased its presence.

Sakharov and dozens of other leading physics were transferred to a secret new complex called Arzamas-16 in a former monastery in Sarov, where they developed the hydrogen bomb. The first Soviet test of a thermonuclear weapon, in January of 1955, killed a number of bystanders kilometers away through radiation or shock wave. Sakharov realized immediately the impact that dozens of these terrible weapons would have. He calculated the non-threshold radiation effects from the repeated—and, he rightfully believed, unnecessary—atmospheric tests, arriving at the figure of 500,000 persons over time. Khrushchev had initiated a moratorium on weapons testing in March of 1958. Sakharov met with Kurchatov, and they agreed that they must convince the Soviet leader of the need for a continued moratorium.[39] Khrushchev warned Kurchatov and Sakharov to keep out of Cold War politics. Throughout this period, it will be recalled, the young Zhores Alferov remained tied to his laboratory—as Sakharov had in his early years—single-mindedly attacking the problem of new semiconductor materials.

Sakharov began to read samizdat literature. His active involvement in the human rights movement commenced with the trial of the authors Andrei Sinyavsky and Yuli Daniel in 1966. Sakharov adopted a public voice in other areas as well. In 1964 he spoke out against the election of Nikolai Nuzhdin, one of Trofim Lysenko's protégés, to membership in the Academy. Sakharov's involvement in the human rights movement crystallized with the Soviet invasion of Czechoslovakia in August of 1968. Brezhnev and his cohort intended to end the Czech experiment with "socialism with a human face." Sakharov expressed his hope over the Czechoslovak reforms in *Reflections on Progress, Peaceful Coexistence and Intellectual Freedom* (April 1968). He argued that world peace required openness, human rights, and engagement so that international security and real disarmament could be achieved. He believed, naively, that there would be convergence between the capitalist and socialist political systems. Over the next fifteen years, until his banishment to Gorky for protesting the Soviet invasion of Afghanistan, Sakharov was a frequent observer at the trials of other dissidents, traveling throughout the nation to monitor them.

Because of Sakharov's activities, leading figures of the scientific establishment joined a slanderous campaign against him—a campaign that Yuri Andropov, head of the KGB and future leader of the Soviet Union, orchestrated. A letter signed by 40 leading academicians, published in *Pravda* on August 29, 1973, called Sakharov ungrateful and implied that he was a traitor. Of the physicists who were full members of the Academy, only Vitalii Ginzburg, Iakov Zeldovich, Petr Kapitsa, and Mikhail Leontovich did not sign.[40] Yet other members of the elite—composers, medical doctors, writers, and film directors, many of whom themselves had suffered in the Stalin era—signed public letters that shared the assessment of the "unworthy conduct" of Sakharov who had "discredited" Soviet science. Common workers and collective farmers joined the campaign.

In October of 1975, Sakharov received the Nobel Peace Prize. This led to another coordinated public attack in the newspapers.[41] Alferov, a corresponding member of the Academy at the time, was not asked to sign the letters against Sakharov. He made himself unavailable in Komarovo at his summer home. He received a call requesting that he join the campaign, but insisted he would not do so. He believes it was "nonsense" for others to have done so. "You can disagree with someone or dislike his political views," he told me, "but good people shouldn't do this kind of thing."[42]

Sakharov's publicly calling the Soviet invasion of Afghanistan a "tragic mistake" was the final straw for the Brezhnev regime. KGB chief Andropov angrily ordered Sakharov's arrest. On January 22, 1980, two KGB officers intercepted him on his way to work and put him on a plane to Gorky, where he spent the next six years in exile. The authorities watched him constantly, bugged his apartment, intercepted all visitors, and monitored his mail. They humiliated him, vandalized his automobile, stole nearly completed versions of his memoirs twice, and denied his second wife, Elena Bonner, appropriate medical treatment. They abused Sakharov and Bonner's immediate families. Three times (in 1981, 1984, and 1985) Sakharov engaged in hunger strikes to protest for human rights and on behalf of his family. The authorities ordered him to be force-fed. Somehow FIAN theoreticians, including Evgenii Feinberg and Vitalii Ginzburg, kept him from being fired from the institute, and they unexpectedly won permission to visit him periodically for scientific discussions.

In December of 1986, Gorbachev ordered the KGB to install a telephone in the Gorky apartment at once. (Citizens had to wait an average of five years for a phone to be installed; in this case Sakharov had waited six years.) When Gorbachev called on December 16, Sakharov pressed for immediate release of all political prisoners. Sakharov returned to Moscow on the morning train on December 22, made his way through a huge crowd of journalists and friends at the station, and went immediately to FIAN, where he spent the day talking science with colleagues. Gorbachev and Sakharov finally met face to face on January 15, 1988. In autumn of that year, the KGB lifted the ban on Sakharov's international travel. Why did Gorbachev wait so long to call Sakharov and release him from exile and to order the KGB to free the scientist to travel? How was this glasnost?

Sakharov's last years were consumed by the dogged pursuit of political reform. In 1989, to force the pace of perestroika against intransigent Party officials, Gorbachev had new election laws passed to rejuvenate the Congress of Peoples' Deputies. Yet the Communist Party held onto power in the Congress through its control of roughly two-thirds of the candidates for office and through the notorious Article 6 of the constitution that guaranteed the "leading role of the Party."[43]

Andrei Sakharov dominated the proceedings at the inaugural Congress of People's Deputies in May and June of 1989. He demanded democratic reform and triggered debate. He forced Gorbachev to permit deputies to give speeches, when Gorbachev clearly intended the proceedings to follow his script. As a dissident, he had a constituency that expected him to hold up the standards of human rights and constitutionalism, although some of his actions disturbed other human rights activists (as when he granted that some individuals who had been subjected to punitive psychiatry actually needed treatment). To the dismay of some environmental activists, he supported an aggressive civilian nuclear power program, with reactors to be build underground to protect them against terror and accidents. But he single-mindedly sought to turn the Congress into a legislative body, used every opportunity to criticize the absence of civil liberties, and angered Gorbachev for taking him to task for attempting to run the Congress from the podium.[44] Still, on many levels Sakharov essentially supported Gorbachev, his pursuit of perestroika, and his foreign policy initiatives, while regretting his reluctance to push reforms to the end.[45]

Gorbachev intended that there be multi-candidate but not multi-party elections. However, democratization led to the formation of scores of unofficial organizations. The Interregional Group of People's Deputies, a parliamentary faction of reform-minded communists belonging to the Congress of People's Deputies and including Sakharov as one of its five chairmen, formed in the summer of 1989 to force the pace of Gorbachev's reforms and to challenge his intention for the Communist Party to retain control over elections by being the only party. The Interregional Group was the closest thing to a new political party. It included almost all of the liberal members of the opposition: Boris Yeltsin, the historian Iuri Afanas'iev, the economist Gavriil Popov, and the Academicians Viktor Pal'm and Andrei Sakharov. For a time, Alferov was a member of this amorphous group. He supported NEP-like market economic reforms, but he did not share the group's demand to abolish legislatively the Communist Party's constitutional monopoly on power.[46]

Alferov and Sakharov were not close professionally or personally, and Alferov did not support the full range of programs of the Interregional Group. He was mostly focused on science. He had a limited but respectful relationship with Sakharov during the latter's short term in the Congress— a relationship that was nearly derailed by another Alferov, one Vladimir Ivanovich, to whom Zhores Ivanovich was not related. When they met in the Parliament just after Sakharov's return from Gorky, Alferov sensed that Sakharov was reserved and perhaps even cold toward him. Alferov believes that this may have been because Sakharov initially believed he was related to Vladimir Ivanovich Alferov,[47] a deputy minister of Minsredmash.

During the spring, summer, and fall of 1989, I had the opportunity to see Zhores Alferov several times in Moscow and Leningrad. He was clearly excited by the events and possibilities of perestroika. Whether meeting me at the presidium of the Academy or the LFTI in Leningrad or for lunch at the Academy's dining hall on Lenin Prospect in Moscow, he exuded confidence that the ongoing political reforms would lead to economic, cultural, and other reforms that would improve scientific performance. He mistrusted market reforms, but he recognized that the Soviet economy could not continue to plod on without stimulus. He also imagined changes in the sphere of education—perhaps something more like the American university system, in which research and education were in concert. As a sign of glasnost, Alferov always had time to talk with me about perestroika, to share recently published articles on the subject, and to point me in the direction of science newsweeklies such as *NTO* and *Poisk*, through which one could follow unfolding debates about the prospects for the scientific enterprise.

Alferov retrospectively characterized his time in the Congress of People's deputies as "dull and gloomy." Educated society had had great expectations for the Congress (which had been hailed as the first democratically elected body since the Constituent Assembly in 1917—itself closed by the Bolsheviks at its first meeting in January of 1918 because the Bolsheviks had not won a majority of seats). Idealists embraced the Congress for its promise of reforms. Many of the deputies were known nationally and internationally. They were, Alferov said, "the gold fund of the country." But according to Alferov it was doomed to failure. No procedures or goals had been prepared.[48] They worked these out on the fly under Gorbachev's chairmanship. There were too many delegates—more than 2,000. It was impossible to accomplish anything with so many voices and in the absence of a political party other than the Communist Party. But it made for great spectacle. Citizens stopped work to watch the sessions of the Congress broadcast live. When tapes of the Congress were shown on television after midnight, people watched into the early morning and came to work, if at all, exhausted.

Science, Advice, and the End of the Gorbachev Era

During the second session of the Congress of People's Deputies, attacks on the Academy began in connection with a growing budget crisis and sky-rocketing inflation. A number of deputies wanted to see the funding provided to the Academy to produce immediate results, and others resented the Academy's wealth. (Beyond its institutes, it had sanitariums, hospitals, apartment buildings, automobiles, clubs, and libraries.) Alferov felt forced to engage in what would become a lifelong campaign to protect the Academy from "unprofessional, faceless bureaucrats."[49]

The Cold War had contributed considerably to the relatively healthy condition of the Academy of Sciences, its institutions, and its members, especially those in the hard sciences; the arms race was good for American and Soviet scientists. Yet during the Gorbachev era a number of scientists and humanists added their voices to a chorus of concern about the danger to science of its heavy reliance on military sources of funding. They lamented what they believed was both insufficient funding and inadequate moral support for basic science, the humanities, and the social sciences. Of course, most Party officials and government personnel considered military and economic applications more important than basic research. At a meeting of the Leningrad Union of Scholars on October 27, 1989, union members discussed precisely this point. They called for a kind of re-education of bureaucrats in the importance of basic research. They sought a transformation of the "stagnant minds" of bureaucrats through the penetration of a scientific worldview into their minds. This worldview included an attitude of respect for fundamental research. They argued that academic freedom was a necessary condition for the existence of science. Unfortunately, as these scholars discovered, a mercenary attitude about the prime economic value of science persisted in government bureaucracies and in the Congress of People's Deputies.[50]

Alferov quickly rose to science's defense as its fortunes declined. He rejected the government's criticism of scientists for not doing more to modernize the economy. Using an example from solid state physics, he pointed out that, if the USSR lagged behind in applications for laser technologies,

that had little to do with the pioneering discoveries of Basov and Prokhorov. Rather, scientists and bureaucrats alike should be grateful that the lag
was not even greater.[51]

On December 16, two days after Sakharov's death, Alferov delivered a
major speech before the Congress on the critical state of fundamental science. Describing the glories of the Academy and noting that it was the
envy of foreign governments and scientists, he warned of the dangers of
even thinking of dismantling it. He had great hope, for he believed the
USSR had begun to transform itself slowly into a better, more democratic
society.[52] But he worried that many of the deputies simply made ignorant
assertions about the state of science. He criticized them for haughty and
even negative comments regarding the Academy's failure to contribute to
the economy. While acknowledging "insufficiencies," he rejected criticisms
that science had done little for the nation in fields ranging from industry to
agriculture. He pointed out that in the nineteenth century Russia had produced such titans of literature as Pushkin, Dostoevsky, Gogol, and Tolstoy,
yet the population had remained illiterate. "But," he asked, "would you
blame the great Russian authors for this state of affairs?" Alferov directed
the Parliament's attention to what he thought were the major problems:
inadequate financing and the wide gulf between science and education.[53]
The older generation had "to pass the baton to the younger," but lacked
the appropriate organizational forms to do so. Alferov called for new kinds
of schools. He worried most about the effort to put science on some kind
of self-financing regimen. He referred to Lenin's aphorism that under communism only three professions would be needed: doctors, teachers, and engineers. To Alferov this underlined the point that health, education, and
science were vital for a society's well-being, and to evaluate them on the
basis of some notion of profitability made no sense. Alferov criticized the
government for its failure to finance science. He provided the example that
the State Committee for Science and Technology set forth fourteen main
directions for scientific research in the nation, with LFTI physicists participating in eight of them, while the government financed only one.[54] Alferov also urged that efforts be focused on cutting-edge research, especially
areas close to his heart and research: information science and technology
and "ecologically pure" energy sources based on microelectronics and
solar cells.

Alferov concluded his speech by criticizing both the far right (namely
the reactionary, anti-Semitic Pamiat) and the far left. Appealing to Leninist ideals, he called for workers, peasants, and members of the intelligentsia to work together to reform the Communist Party. He referred to his

working-class background to remind the other deputies of the potential of socialism. "The fate of perestroika," he concluded, "is in our hands."[55]

Perhaps one of the failings of the Gorbachev administration was its inattention to science policy. The government had created a State Committee on Science and Technology in 1961 to coordinate R&D, and relied on a variety of Academy of Science commissions to assist it in formulating policy. The Central Committee of the Communist Party had a secretary responsible for issues of science. But there was no position comparable to the Special Assistant to the President (of the United States) for Science, nor was there an institution comparable to the President's Science Advisory Committee. Gorbachev seemed to want both something more informal (colleagues who were specialists in pressing issues of science) and something more formal (perhaps a Soviet equivalent of the President's Science Advisory Committee), but nothing ever came of this. I had the sense that Alferov was one of Gorbachev's informal science advisors. "We weren't advisors," Alferov told me. "He asked questions of us. Instead he had a president's council with [Yurii] Ossipian [a solid state physicist with interest in arms control], [Stanislav] Shatalin [an economist whose served in GKNT and the Academy of Sciences], and [Nikolai] Petrakov [a specialist in economics who allied with Shatalin in the 500-day "shock therapy" economic reform plan, a plan eventually rejected by Gorbachev]."[56]

Other observers believe that Evgenii Pavlovich Velikhov, head of the Kurchatov Institute for Atomic Energy, had a greater role in advising Gorbachev directly and had greater access to the general secretary than other scientists had at any other time in Soviet history. Velikhov—a plasma physicist, like Alferov a vice president of the Academy of Sciences, and like Alferov a man of expansive scientific vision—led Academy efforts to promote "computerization" of the USSR. The effort fell well short because its top-down, centralized approach significantly slowed broad-based social acceptance and the development of computer culture. In the 1980s, Velikhov provided expertise to Gorbachev on disarmament issues.

Gorbachev's "new thinking" required new approaches to arms control. In 1983, as Frank von Hippel recalls, Federation of American Scientists officials and staff met with the new Committee of Soviet Scientists led by Velikhov. "Velikhov promoted Gorbachev's 1985–1987 unilateral nuclear test moratorium and obtained permission for the Natural Resources Defense Council (NRDC) to install seismometers around the Soviet test site—the first time the previously paranoid Soviet leadership had accepted in-country verification." The Federation and the Committee carried out joint studies and demonstrations to demonstrate both the promise of arms treaties and the dangers of a continued nuclear buildup.[57]

The historian Vlad Zubok recounts that Gorbachev advanced the change in policy from "from military preparedness to domestic reforms and disarmament, from orthodoxy to new thinking" in a speech at the Party Congress in February of 1986. In late May, in a secret speech to the senior personnel of the Ministry of Foreign Affairs, Gorbachev announced: "Peace is the value above anything. In the nuclear-cum-space era a world war is the absolute evil. It cannot be won, as well as the arms race. . . . The threat of nuclear war cannot be ignored when one discusses the prospects of world class struggle." At this time, the Velikhov commission was considering the destabilizing nature of Ronald Reagan's Strategic Defense Initiative, and whether to recommend a response with financing for a Soviet crash program. Scientists and policy makers presented Gorbachev with a dozen options. Gorbachev, a lawyer by training, lacked the scientific expertise to evaluate them. Though he never took a firm stand on a Soviet SDI, the debate alerted him even more to the need for arms-control endeavors. The Chernobyl disaster finally convinced Gorbachev of the need to "to look at the task of nuclear disarmament as a moral imperative independent of political calculations." Velikhov was the head of the commission appointed by the Politburo to report on Chernobyl's causes and significance.[58]

Another of Gorbachev's advisors was Roald Sagdeev, whose marriage to Susan Eisenhower (a granddaughter of the former American president) suggested a new form of high-level Soviet-American détente. Sagdeev advised Gorbachev on issues related to civilian space and space-based weapons systems. He had been a wunderkind in plasma physics, first at the Kurchatov Institute, then at the Institute of Nuclear Research in Akademgorodok. In 1973 he was named director of the Institute for Space Research. Elected to the Congress of People's Deputies from the Academy slate, along with Alferov, Sakharov, and others, he often sided with the Interregional Group on reforms.[59]

· Alferov's science advisory activities—aside from the occasional remark or hallway discussion—were largely limited to his service on the Congress's Committee on Science and Education (and later on the same committee in the Russian Duma under the chairmanship of Nikolai Gregorievich Malyshev). From this position, Alferov indicated a strong belief in the sanctity of scientific truth and its ability to operate independently of political pressures. But the reality was more complex. First, the deputies were engaged in constant discussions over the future of basic science and higher education, finance, the potential role of grants, and the proper level of accountability of scientists to state funding agencies. Clearly there was more to science than the specialist seeking to establish the truth. Second, Alferov enunciated the need for a kind of scientific responsibility; he believed complete glasnost engendered scientific responsibility that had been lacking during the preceding 60 years when the Soviet leadership and engineers pushed big projects. Glasnost led to the publication of exposes on the costs of the Soviet development paradigm. These costs went well beyond financial costs to human and environmental costs, from Chernobyl to the highly polluting chemical and metallurgical industries, and to reckless nature transformation projects—dam, irrigation, and water transfer systems.

Though clearly a member of the establishment, Alferov was critical of the role of big science and technology, and particularly of the rush to employ various engineering strategies to rebuild nature according to Soviet ideological and planning desiderata. He believed that the desertification of the Aral Sea region for agricultural purposes and the construction of paper mills on the shores of Lake Baikal indicated that scientists had permitted dangerous projects to move ahead with inadequate consideration of all costs. Lake Baikal—the "jewel" of Siberia and the largest freshwater lake in the world by volume—had fallen to military and pulp and paper engineers. To avoid such projects in the future, Alferov called for "complete glasnost" in the scientific sphere.[60]

Like many specialists, Alferov voiced his concerns about big science and technology publicly only during the Gorbachev era. Referring to a controversial 25-kilometer-long dam complex on the Neva Bay touted for flood control, Alferov said: "I cannot judge whether a dam is needed—in this area I am not sufficiently competent. But I cannot agree with the arguments of the proponents of dams: 'We have gotten 300 or 500 impact statements from different instances.' The collection of statements—this is not glasnost. Huge projects—and this regards not only those cases when we invade nature—should be discussed openly." Perhaps his lifelong focus on the microworld engendered his skepticism. Alferov came to believe that the larger and more expensive the project, the greater the responsibility of its champions to discuss its merits and true costs openly. He recognized the nearly universal tendency for a project to acquire nearly unstoppable momentum. Even when a project turned out to be a mistake, the government continued to fund it. This type of technological momentum was common in the USSR.[61] The Neva dam complex, with construction on hold since the fall of the USSR and an ever-ballooning budget, gained the strong support of President Vladimir Putin, another Petersburg patriot.

Ultimately, Gorbachev, reformers, conservatives, and other intellectuals were too consumed by politics to pay attention to Alferov's beloved science. Citizens had less concern about science after the economy began a downturn with accelerating inflation that would be reversed only in the late 1990s, well after the breakup of the USSR. Beginning in the 1990s, any welcome reforms in science were ultimately overshadowed by grave new problems that institute directors and national administrators like Alferov were powerless to combat. The first was a brain drain. Leading specialists, especially theoreticians, accepted one-year or longer postings at institutes abroad, and many of them never returned to Russia. Even more significant was the internal brain drain that occurred when young students abandoned science for business or other fields. At the same time, the average age of scientists increased rapidly, as did the number of scholars earning Doctor of Science degrees. (The latter was attributable to greater ease in defenses.) Skyrocketing inflation removed whatever cushion remained. Many institutes had to operate without heat or electricity for long stretches since they were unable to pay their utility bills. Funding from European and American organizations and from the Hungarian-born financier George Soros in the form of grants and contracts through his International Science Foundation was of crucial importance in this time of economic uncertainty. Yet here remnants of Soviet thinking interfered. In many cases institute directors insisted that all announcements of grant opportunities be made through their offices, and they shared information about foundations only with select, favorite colleagues. Finally, the effort to establish a more open and fair system of peer review to allocate funds among competing projects encountered resistance from older scholars.

Unsettled by the decline of Soviet science even before the breakup of the USSR, troubled by its loss of funding and prestige, and bothered that parasitic cooperatives were eating away at the Academy from within institute walls, Alferov sought to use his various administrative positions to save basic science from collapse. He had welcomed Mikhail Gorbachev and perestroika. Since the reform movement Gorbachev set in motion was based on open and honest assessments, Alferov figured it must be good for science. Instead, perestroika triggered processes that threatened the scientific enterprise. If he had any hopes that the Yeltsin presidency might contribute to the stability of science, those hopes were dashed in the political crises and steep economic decline that accompanied the creation of the Russia Federation out of the rubble of the USSR. Alferov would have to use all of his energy to protect science from further losses.

6   Scholar, Laureate, and Statesman

Russians play a high-stakes game of politics. They have a lot to play hard about. Russia's natural resources of oil, gas, minerals, and timber are perhaps the richest in the world, and since the breakup of the USSR businessmen and politicians have fought over the disposition of those resources. Many of these individuals are former Communist Party officials who learned to play politics in the Soviet era, and they often use tactics from that era. They have their sights on real estate, too; the booming real estate markets of Moscow and St. Petersburg are central arenas of a struggle over Russia's great wealth, and the fight over ownership includes Russian Academy of Sciences (RAN) institutes, clinics, housing complexes, hotels resorts, automobiles, and other assets. Former president Vladimir Putin, current president Dmitri Medvedev, and Minister of Science and Education Andrei Fursenko believe that the RAN in its present form is not sufficiently accountable to the state, and that funding levels ought to reflect research of more direct benefit to the economy and government. They, and other critics, wish to reform the RAN's administration, push its leadership to be more responsive and flexible, see applied research take a rightful place in institute programs, and tie the RAN generally to larger overarching national programs directions of research framed by the ministry. In their book *Science in the New Russia*, Loren Graham and Irina Dezhina refer to the ongoing efforts to define the status of the Academy as a network of research institutes with government and foundation funding, rather than a self-governing honorary society that elects in own members and officers, as the "governmentalization" of science. But they point out that what has happened, what will happen, and what the status of the RAN will be all remain unclear.[1] In any event, the Medvedev government wants greater accountability for its funding—and eyes the RAN's vast real estate and movable property resources with great interest.

Since the collapse of the Soviet Union, the scientific establishment has passed through a series of crises. The crises were connected with a sharp decline in the level of financing that called forth the involvement of foreign governments and foundations to salvage the enterprise. That involvement occasionally provoked xenophobic concerns that foreigners were stealing the scientific legacy. Another problem was the "brain drain" of young persons to other professions at home and of more established scholars to research centers abroad. Finally, conflict arose between members of the prestigious Russian Academy of Sciences, the locus of basic research, and a federal government determined to rein in its independence, and perhaps to gain control of its vast resources. But there is no question that the enterprise begged for reform. "Rather than being the best science in the world," Graham and Dezhina write, "science in the late Soviet Union was crying out for reform. It was a system that emphasized quantity over quality, seniority over creativity, military security over domestic welfare, and orthodoxy over freedom."[2] Zhores Alferov has been involved directly in policy debates over the disposition of the post-Soviet scientific enterprise.

Putin, although according to polls one of the most popular figures in Russian politics, used his authority and the increasing power he acquired to enable the government to re-take control of the nation's wealth, including natural resources and real estate. The government has used media that it dominates to frame debates and limit dissent. Alferov, as a parliamentary deputy who represents the Communist fraction, a man of great authority as a Nobel laureate with seemingly unassailable connections in the Academy's ruling structure, has entered the power struggle to determine what direction the Academy's research programs will take and to whom its vast resources belong. He struggles to protect basic research, which he sees as critical to Russia's future. He believes that an effort to set up a number of elite national laboratories on the basis of Academy institutes will serve the government's goal of contracting the Academy all too well. In the process, the government will complete the devastation of a world-class R&D enterprise dedicated to basic science. His fight to establish a new scientific research and educational center (Nauchno-Obrazovatel'nyi Tsenter, or NOTs) to bring higher education and basic science together under one roof continues against long odds. It is a battle that he says has caused him "great pain and suffering" while leading him to distance himself from the Ioffe Institute, to which he has devoted the last 50 years of his life. Alferov's teaching and research activities have also drawn him away from LFTI and toward NOTs.

In 1991, Alferov was in a position, as director of LFTI , vice president of the Academy of Sciences, and a member of the Parliament, to build on the reforms of the late 1980s to improve the performance of science. But the obstacles to reforms were daunting. After the breakup of the USSR, Alferov and other scientific leaders faced a precipitous decline in support of science that prevented any effective reforms. Instead, over the next decade, Alferov was forced to use his various positions as statesman and administrator to try to save science from utter catastrophe. Junior scientific talent left the LFTI and other institutes in droves (many for business; others, especially theoreticians, for short-term employment abroad), and a large number of them would not return. New laws concerning grants and contracts, especially from foreign sources, undermined what little financial security institutes had, since the government insisted on taxing the grants, and the banks that arose from the rubble of the USSR took an obscene share of the funds as processing fees. Officials in the executive branch and the Parliament focused exclusively on the political and economic controversies of the day, including hyperinflation and Russia's decline as an international power, not on education, environment, public health, or other such issues. The result was the near destruction of fundamental research. The latter was saved only by the intervention of a variety of foreign governmental and private endeavors, with the most publicized of the private endeavors underwritten by the financier George Soros.

At its nadir, late in 1993, financing for science was perhaps one-twentieth of its Soviet level. In an interview published in the *Neva Times* on September 18, 1993, Alferov lamented the state of his institute and of fundamental research generally. The USSR had been a leader in many fields, but now several major scientific schools were on the edge of collapse. Young people hesitated to choose science as a career, and talented researchers bolted from their institutes at the first opportunity. The Yeltsin administration managed to find some financing for military research and development, but, Alferov asserted, confused support for the military-industrial complex and its important contributions to innovation and economic growth with support for basic research, for only 5–10 percent of science in the military-industrial complex was real (by which he meant basic) science. On the other hand, perhaps 75 percent of the research conducted in the entire system of Academy institutes, universities, and industrial ministries was military.[3] Alferov continued by severely criticizing new tax laws that in essence confiscated up to half of the funds awarded in grants to an institute, a laboratory, or a scholar. For two years the LFTI had been unable to purchase new equipment in such miserly circumstances. At the LFTI, as at other institutes, the directors resolved to pay salaries to staff members rather than fire redundant personnel and use the kopeks remaining to keep some semblance of research going; virtually all moneys went to salaries. Alferov strove to remain optimistic about the future, primarily because the quality of researchers remained high. But if the situation did not change shortly, he had no idea what might occur: "The matter is not only in low salaries, but the very possibility of doing science at all in our country."[4]

During the Yeltsin years, Alferov served on various committees in the Parliament in the attempt to save the legacy of Soviet science for the future of Russia. Finding a lack of interest or a lack of knowledge among government officials, and a lack of cooperation among his fellow deputies, Alferov turned to the Communist fraction. Remember that his father was a convinced Communist who had met Lenin and Trotsky and deeply respected the latter, that his uncle also was a member of the Party, and that Alferov had joined the Party not only because it was expected among leading scientists but also because he shared the values of the Party leadership about public health, higher education, and science (at least the values expressed by its reformist wing). He had seen the Gorbachev reforms as holding promise for science, and had joined the Interregional Group of Parliament deputies, which included Boris Yeltsin and Andrei Sakharov, to support those reforms. But as science declined, Alferov grew increasingly critical of post-Soviet leaders for their failure to protect basic research.

Alferov determined to use his positions as a leading scientist, administrator, and parliamentary deputy to defend the scientific enterprise. He served on a parliamentary committee responsible for education and on another concerned with the disposition of spent nuclear fuel. But he has mostly used his positions and public persona to make the case to support basic research before a government which he believes has only a mechanical and superficial understanding of its importance to the future of the nation. He knows Putin (now the prime minister) personally, and on many levels respects his accomplishments, yet he criticizes Putin's unwillingness to instruct various ministries to increase funding for science and higher education dramatically.

Alferov sees himself as a savior of science, and he sees St. Petersburg, with its legacy as the crucible of Russian science, as the stage on which he must carry out the struggle. Peter the Great brought science, with its Western world view and its secular educational institutions, to Russia in the early eighteenth century, but did not live to see the Academy of Sciences open officially. (It opened in 1725.) Petersburg was home to Soviet science until Stalin moved its presidium to Moscow in 1934. Now Alferov believes that, on the foundation of the Academy's St. Petersburg Science Center, with its direct lines to Peter, and with its leading physics, chemistry, and biology institutes, he must pursue new programs to revitalize basic research. He sees a new university and a new lyceum as the institutional keys to the future, and nanotechnology as the foundation of twenty-first-century science.

As an internationally renowned scholar with access to persons in government and to such resources as grants, contracts, and the budget process, Alferov sought to bring the strengths of Soviet science and Western organizational ideas together. A concrete example of this union is NOTs, located only an eight-minute walk from the LFTI. Having retired as director of the LFTI, Alferov uses NOTs to promote a kind of liberal arts approach to education, combining an emphasis on the sciences with classical training in the humanities, languages, and physical education, to identify and train high school students for careers in research, including research in the field of nanotechnology. He travels widely within Russia to argue for support of science, and travels even more widely abroad giving talks on the importance of nanotechnology in modern economies. He has offended his colleagues at the LFTI by pushing for the use of its resources to support his vision of Russian science. He has angered government officials with his incessant criticism of their failure to embrace his vision. Above all else, he remains frustrated that the glorious Soviet legacy of theoretical physics, nuclear physics, solid state physics, and heterostructures has been diminished by the failed policies of the Yeltsin and Putin administrations.

The Yeltsin Presidency and Science

Whether from inattention and incompetence, greater concern over holding on to power, inability to keep the Russian economy from falling into a deep recession in the early 1990s, or a combination of the three, President Boris Yeltsin, whose administration took power with great hopes of building a democratic future on the foundation of a market economy, did nothing to support fundamental science in a time of crisis as great as that just after the Russian Revolution in 1917 and 1918. Even then, scientists managed to expand the scientific enterprise, not watch its collapse.

Boris Yeltsin, grandson of a de-kulakized peasant and son of a worker who served three years in the gulag, graduated from the Ural Polytechnical Institute with a degree in construction engineering in 1955 and joined the Uraltiaztrubstroi Construction Company in Sverdlovsk. He became a member of the Communist Party in 1961. Later that year he took a job as head of an industrial apartment building construction trust. Ascending rapidly through the Party apparatus, in 1976 he was appointed secretary of the Sverdlovsk Provincial Party Committee. In 1977 he followed orders to destroy the house in Ekaterinburg (then still Sverdlovsk) in which the tsar and his family had been executed. In April of 1985, Yeltsin moved to Moscow to various positions in the construction industry, rising to First Secretary of the Party's city committee. He ordered the renovation of Arbat Street into a pedestrian mall. He was then elected to the Politburo in the Gorbachev administration. After criticizing the slow pace of reform at the October 1987 plenary meeting of the Central Committee, he lost his positions as Moscow Party secretary and in the Politburo and was demoted to First Vice Chairman of the State Committee on Construction.[5]

In March of 1989, thanks to the popularity he had achieved for ordering the renovation of Arbat Street and the notoriety he had gained for criticizing the pace of reform, Yeltsin was elected to Congress of People's Deputies as a deputy from Moscow's Electoral District No. 1. One of the more populist positions of his electoral program was a call for reduction in spending on the space program. From the Congress he was elected to the Supreme Soviet of the USSR, where he held the position of chairman of the Committee on Construction and became a co-leader of the Interregional Group of deputies, and in a series of public forums (one of which I had the good luck to attend) captivated public attention with his insistent explanations of the importance of radical perestroika. In May of 1990 he was elected speaker of the Supreme Soviet of the Russian Republic. From that position he intensified his criticism of Gorbachev for the slow pace of reform, for failing to deal with the Party conservatives more firmly, and for refusing to recognize the need to allow other political parties to form. On June 12, 1990, the Russian Congress of People's Deputies adopted a Declaration of Sovereignty, which accelerated the breakup of the USSR. Shortly thereafter, Yeltsin quit the Communist Party.

On June 12, 1991, Russia's first democratic presidential election was held. Not surprisingly, Yeltsin won. Gorbachev and the heads of the Soviet republics, including Russia, were scheduled to sign the Union Treaty, which would hold the USSR together as a series of independent states, on August 20. To head this off, Communist hard liners attempted a coup on August 19. They detained Gorbachev, who was on vacation in Crimea. Yeltsin—against the wishes of some of his advisers, who feared that he too would be arrested—rushed from his home on the outskirts of Moscow to the White House (the Parliament building). The conspirators had ordered tanks to surround the White House. Thousands of unarmed citizens came to defend it, and one of the tank units joined them. Yeltsin climbed atop a tank, condemned the coup, and called for resistance. The coup failed when its leaders lost their nerve in the face of this public display. (Alferov told me that he believes the coup's leaders were attempting to defend Soviet law, not engaging in illegal activity.) On December 1, Ukraine held a referendum and its citizens voted for independence from the Soviet Union. A week later, the presidents of Russia, Ukraine, and Belarus signed a treaty creating a Commonwealth of Independent States. On December 24, Russia took over the USSR's seat in the United Nations. The next day, Gorbachev resigned. The Soviet Union ceased to exist, and Yeltsin became the president of the largest nation of the former Soviet Union, Russia. Fortunately for non-proliferation, Russia also became the inheritor of the empire's nuclear weapons when Belarus, Kazakhstan, and Ukraine willingly gave up their weapons to it.[6]

Economic crisis, political turmoil, and even bloodshed characterized the roughly eight years of the Yeltsin presidency. As his charisma faded, Yeltsin also faced charges of cronyism, nepotism, and corruption, and many citizens ridiculed his apparent love for alcohol. Yeltsin's first government attempted to introduce a program of liberalization of prices, legalization of private business and private ownership of land, introduction of free trade and commercial banking, massive privatization of state-run enterprises, and radical cuts in military spending.[7] But recession and inflation destroyed the earnings and savings of the average citizen, while well-placed former Communist officials managed to strip many of the important state-owned assets during a period of violent privatization. Domestic production fell, and organized criminal groups moved in to control what they could.

In September of 1993, Yeltsin dissolved the Russian Parliament, declaring that the opposition of its large number of Communist holdovers had paralyzed his reforms, and announced new elections. Yeltsin may have felt insulted by a member of the opposition who made a gesture indicating that Yeltsin was a drunk, but there is evidence that the Parliament's leaders intended to remove him under the old Soviet constitution and appoint one of their own. The Parliament responded by voting to depose him. Yeltsin ordered the police to surround the Parliament and cut off the electricity. In October, supporters of the Parliament broke through police lines and rampaged through Moscow. Yeltsin, at his dacha in Peredelkino (a prestigious village where many of Russia's intellectual and political elite have lived), panicked. He called in troops and tanks. The White House was shelled and left in flames, and hundreds were killed. When he returned to the Kremlin, Yeltsin imposed an overnight curfew and banned extremist opposition parties. He briefly closed down *Pravda* and other newspapers that had supported the rebels, and he jailed the opposition leaders (who were later pardoned by the new Parliament). Over the next three years, the Russian government accomplished little in the way of domestic economic or political reforms, and the fortunes of scientists continued to plummet.

In December of 1994, hoping to end flaring independence movements, secure oil and pipeline routes in the Caspian Sea region, and improve his popularity, Yeltsin ordered an invasion of Chechnya. In the preceding three years, civil war had broken out there after a declaration of independence. Russia had supported opposition groups clandestinely. But after the invasion a full-scale guerilla war ensued. Tens of thousands of Russian and Chechen soldiers and perhaps 100,000 civilians have perished since then, and the capital of Chechnya, Groznyi, has been reduced to rubble.[8]

By 1996, the Communist opposition had grown significantly owing to continued economic, political, and military failings. Yeltsin announced his candidacy for a second term to prevent a Communist candidate from gaining the presidency. Yeltsin won the election fairly handily, and the next four years were marked by greater stability, although his presidency was plagued by charges of nepotism and corruption, and the economy did not begin to recover until the collapse of the ruble in August of 1998.

When the ruble lost more than 80 percent of its value relative to foreign currency overnight, most people anticipated acceleration of Russia's downward spiral. But since the ruble was now worthless against the dollar and euro, imported goods were too expensive. Russian industry and agriculture surged forward to make up the difference, and the economy grew rapidly on the basis of the sale of natural resources, notably oil, gas, forestry products, nickel, and copper.

When the USSR disappeared, many of its resources became the property of the Russian Federation. How the Yeltsin administration dealt with those resources affected science. During the Soviet period, scientists had gained permission to establish an Academy of Sciences in each republic. With the dissolution of the USSR, each republic became a nation, and each republic's academy of sciences served as the foundation of a new national academy of sciences. The leadership continuity in many of these academies has surprised some observers; in the extreme case, the president of the Academy of Science of Ukraine, Boris Paton, first appointed in 1962, remains president in 2009. In the confusion of the breakup of the empire, a number of scientists in Russia determined to create a Russian Academy of Sciences, and even gained a charter from the government to do so. Many of the individuals working in academy institutes felt slighted because they were not full or corresponding members of the prestigious organization. Members of the Soviet Academy who were residents of Russia opposed the move. Younger scientists criticized the entire system, in which a kind of gerontocracy of several hundred individuals determined the direction of basic science research for the entire nation. More democratically inclined parliamentary deputies, for their part, were critical of the Academy's "unaccountability" to the government. Rather quickly, however, by the end of 1991, the Russian Academy and the previous Soviet Academy were integrated, and new elections saw a somewhat wider and younger membership.[9] The Yeltsin years were a catastrophe for Alferov's LFTI and similar Academy institutes.

The Pauperization of Russian Science

Though the Yeltsin administration saved the Academy of Sciences from juridical and organizational collapse, it did not address the financial problems that accompanied the inflation of the 1990s. On top of this, the share of federal budget for science fell significantly, leaving Russia far behind other leading scientific countries (the United States, France, Britain, Japan) in percentage share of federal budget or GDP. Whereas in 1991 total expenditures on R&D from the federal budget and all other sources amounted to 4.28 percent of GDP, in 1995 it had declined to 1.39 percent of GDP; in 1998 it was 1.33 percent; in 2001 it was 1.79 percent, and in 2003 it was 1.96 percent. The number of researchers per 10,000 of population also dropped precipitously from 1991 to 1995 as many scientists left the field and young people decided not to enter graduate school, although the number of scientists then recovered and held stable.[10] But the decline in funding far outstripped the decline in personnel. This meant that available funds went primarily to salaries and wages, not to research or overhead. A long-awaited federal law "On Science and Government Science and Technology Policy" came into force in August of 1996. According to the statute, science was pegged at 4 percent of the federal budget. The government subsequently issued a series of declarations that showed concern about science, but financing came nowhere close to statutory promises.[11]

Alferov and his colleagues in the St. Petersburg (formerly Leningrad) Scientific Center worked vigorously to secure funding. In the 1990s the center sponsored a series of talks on "The Experience of Work of Directors of Academy Institutes in a Crisis Situation" to draw attention to science's deepening plight. For example, on March 26, 1992, at a joint meeting of the center's presidium and the St. Petersburg mayor's council on science, and higher and middle education, Alferov spoke about the great scientific potential in St. Petersburg's institutes, museums, and archives. Yet economic crisis and political instability had relegated institutes to the position of beggars. Appeals to city, regional, and national government got nowhere, because officials were not focusing on issues of science. In 1994 the mayor of St. Petersburg established a formal department of science and higher education, but it is not clear how a municipal government anywhere in the world could support scores of research programs, let alone this government (which could not repair sidewalks, patch potholes, or restore public transportation). The federal Ministry of Science and Technology Policy agreed to include funds for science in the department's share of the municipal budget. The total sum was pegged at 0.11 percent of the budget, but little more than half of that minuscule amount made it to the institutes, of which one-seventh went to research.[12] A number of institutes hung over the abyss of closure (including the Academy's archive, whose employees left in large numbers).[13]

This situation was repeated in all institutes, including the crown jewel of St. Petersburg science, the LFTI. On December 3, 1993, several major local and national newspapers published an open letter from scientists at the LFTI to Deputy Prime Minister Egor Timurovich Gaidar. Gaidar, a graduate of Moscow State University, an editor of the Communist Party's leading theoretical journal, *Kommunist*, and a former prime minister, was one of the authors of a plan to use "shock therapy"—rapid privatization of state industry and monetization of the economy—to kill the Soviet planning system during the first years of the Yeltsin administration. Shock therapy killed not only the planning system but the economy. Production dropped precipitously. Inflation reached 2,500 percent. As a result, in real terms the monthly budget for the LFTI fell by four-fifths in the first nine months of 1993. Total official funding had increased 150 percent, but with inflation four times greater and the salary indices up 360 percent the result was a miserly budget for equipment and reagents. By October, funding for salaries had disappeared. Many institutes throughout the nation had no money for phones, electricity, or heat. Botanical gardens and biology institutes throughout the nation lost irreplaceable specimens. To make ends meet, people sold everything they could at flea markets. Some even brought burned-out light bulbs to their workplaces, pilfered the good bulbs, and installed the non-working bulbs in their stead.[14] In this environment, science was impossible.

The 235 LFTI signatories of the open letter to Gaidar included 150 doctors of science and candidates of science. Naively claiming to steer clear of politics, they called for a change in government policy and for emergency infusions of funding to science. They referred to the preceding two years of radical reform under Gaidar's direction not, they wrote, to address matters of the successes or failures of reform, for that was a matter for politicians and economists to consider, but rather to bring the critical state of fundamental science to the attention of the Yeltsin administration. They rejected the prevailing view among politicians that science was irrelevant to economic growth; they believed government policies toward academic institutions demonstrated precisely a lack of interest in and a lack of understanding of science. The scientists reminded Gaidar that the performance of Russian science was equal to that of other leading scientific nations and had contributed to the nation's great achievements. But to maintain this performance, they needed funding. They pointed out that, at best, scientists' salaries were perhaps one-half of the national average; in reality they had lost virtually all of their purchasing power.) A recently passed law that gave scientists merit increases for advanced degrees and other achievements meant nothing even if prices were indexed to inflation because of the absence of funds. The letter concluded with a complaint that the government had spent one-and-a-half times the Academy of Sciences' annual budget on renovations to the Parliament Building.

A political cartoon that appeared in the newspaper *Vechernii Peterburg* (*Evening Peterburg*) in February of 1993 indicated how far science had fallen in terms of prestige and government support. Two beggars sit on the ground with their hats upside down in front of them, pleading for spare change. A third beggar has written a series of equations on a wall. One of the other two beggars says "He's been assigned to us from the Fiztech."[15]

Out of concern for colleagues, and for a series of other reasons, a wide range of scientists, philanthropists, and policy makers launched an international effort to provide emergency funding to scientists in the former Soviet Union. Many people who supported this effort recognized that the tradition of world-class performance in a variety of fields was reason enough to provide funding. Many others were worried that scientists with knowledge of nuclear and other weapons might be tempted to offer their services to Iran, Iraq, North Korea, and other "rogue nations." The governments of the European Union countries established the International Association for the Promotion of Cooperation with Scientists from the former Soviet Union (INTAS) to develop a support system.[16] The International Science and Technology Centers (ISTC) grew out of the efforts of specialists in the United States, the European Union, in Japan, and at CERN (the European Center for Nuclear Research) to provide funding for programs to keep individuals with classified knowledge at work in Russian research institutes.[17] And the

International Science Foundation (ISF), funded by George Soros (a Hungarian expatriate dedicated to the creation of democratic institutions in the former socialist countries), came to the rescue of many former Soviet scientists.[18] The ISF will serve as one example of the challenges and successes of international rescue operations for science in the former Soviet Union.

Soros announced the creation of the ISF in December of 1992 with a $100 million gift to support emergency individual grants of $500 each. Boris Saltykov, minister of the Russian Ministry of Science and Technology, publicly endorsed the ISF. In private meetings he emphasized the importance of giving money to institutes somehow, for example through overhead or equipment purchases, out of fairness, to prevent jealousies from developing among staffers who received grants and those who did not, and to head off the effort of institute directors to confiscate some of the $500. Quickly a network of regional offices of the ISF spread throughout the former Soviet Union. Although the ISF managed to get grant programs running quickly and thereby to save many a scientist from ruin, a growing chorus of Russian nationalists and other conservatives commenced a campaign against it and other Western foundations with the accusation that they were buying scientific secrets and personnel on the cheap. The successor to the KGB, the FSB (Federal Agency for Domestic Security), conducted an investigation and reached the inflammatory and false conclusion that a situation close to espionage existed, and that "brain theft" had occurred. Other opponents worried about the theft of industrial secrets at the behest of the American Central Intelligence Agency. Soros publicly stated he would cease operations if the government did not officially approve the ISF's activities as spelled out in its charter, a document of measured tone and openness that should have ended any doubts that the ISF intended to support basic, non-classified research in Russia. Ultimately, parliamentary hearings on the matter led the Yeltsin government to welcome the ISF's support (which clearly had nothing to do with espionage) and to stop the FSB's interference.

The ISF offered emergency grants of $500 to scientists who had published at least three articles in the leading open-source scientific journals in the preceding five years. It also provided $1,500 grants to 1,000 different research groups, and 274 grants of $1,500 to institutes for equipment purchases. More than 26,000 scientists in the former Soviet Union, including 20,763 Russian citizens, received $500 each. In some cases they encountered difficulties in the disbursement of funds because outdated laws forbade certain kinds of transactions in foreign currencies, and the Russian banking system could not handle transfers expeditiously. The banks

intended to gouge grantees of 30 percent of their grant funds through ne-
farious processing fees. A library assistance program gave some 500 libraries
approximately 130 different foreign journal titles, plus textbooks. And a
conference travel grant program that in 1996 got an infusion of funds from
the Russian oligarch Boris Berezovsky enabled 10,000 scientists take part in
more than 2,000 conferences.

The way ISF funds were disbursed embarrassed many scientists. According to one newspaper report, in November of 1993 a long line of people
in disheveled clothing, old jackets, and scuffed shoes appeared at the St.
Petersburg Innovation Bank to receive their $500 grants. They stamped
their feet in the cold to keep warm. Among them was Dmitrii Balin of
the Gatchina-based Institute of Nuclear Physics, a specialist in molecular
nuclear physics and a father of three, who said he welcomed the money,
since he couldn't live on his salary of 30,000 rubles ($30) a month. One
scientist used his $500 to buy a ticket to Milan, where he worked in an
Italian institute under contract; another bought furniture for his new apartment. Others had work done on their summer homes. One man filled out
the required grant application forms but worried that the Americans were
purchasing scientific knowledge on the cheap through those very forms. A
physicist from the LFTI found the entire process uncomfortable. He complained that the grants were organized "in a Soviet fashion." First, he had
to stand in line for an hour in the freezing cold. Then "inside the bank
the . . . security police and tellers treat us as if they don't want us here." Six
hundred LFTI researchers earned grants. Though $500 was a small amount
to Western scholars, many of whom made $5,000 to $8,000 a month, it
was crucial support for scholars trying to get by on less than $50 a month.[19]

Although Zhores Alferov questioned some of George Soros's motives in
establishing the ISF, he thanked him for his efforts to save Russian science
from financial ruin in the early 1990s. Alferov believed that Soros was not
entirely altruistic, certainly seeking personal recognition as much as having philanthropic goals in mind. After all, to Soros "$100 million is not a
lot of money." But the assertion that Soros wanted to buy Russian secrets
on the cheap, or to control the direction of the development of science, as
some nationalists claimed, was "utter nonsense" according to Alferov. He
contended that the $500 emergency grants to scholars were an important
gesture if not crucial to the rescuing of the scientific establishment. The
LFTI's budget had been roughly $75 million in 1990; by 1992, as a result
of budget deficits and inflation, it had shrunk to $4 million in real terms.
Beyond the emergency grants, LFTI scientists received 77 Soros research
grants totaling $2 million.[20] That $2 million saved the LFTI, Alferov said.

Another achievement of Soros was that "he taught us how to compete for grants."[21]

An important aspect of the changes in the management of science after the breakup of the USSR—promoted to a large extent by the appearance of grant and foundation money—was the introduction of peer review to determine project funding. In the past, institute and laboratory directors had largely distributed annual line-item block funding to the programs and individuals of their choice. Now competitive review of research projects became more widespread as a way to determine which projects merited funding. Although the amounts of funding from foreign sources and from the newly established Russian Foundation for Basic Research (Rossiiskii Fond Fundamental'nykh Issledovanii) remained small in the 1990s, the creation of a peer review grant system encouraged greater reliance on originality and meritoriousness as criteria to select projects as opposed to connections with leading scientists or institute and lab directors' desires. Many Russian scientists now accept and indeed prefer a system of funding based on peer review, even if they worry about the low levels of funding.[22]

The creation of the Russian Foundation for Basic Research was an attempt to modernize methods of funding science and initially to ease deep financial crisis. Rather than have rotating chairs of various programs, as the (US) National Science Foundation does, the Russian Foundation for Basic Research has permanent personnel, many of the based in Moscow. Some worry that this may lead to geographic favoritism. Yet the RFBR has contributed to greater openness and competition, in contrast with the old system of big institutes dominated by scientific administrators. Scientists view the funding as still inadequate, but it has increased significantly (from 1.73 billion rubles in 2002 to 2.84 billion rubles in 2006). The RFBR seems to strive to fund a larger number of smaller projects to spread funding wider with the result that each project receives less money than perhaps it ought to, but at least research moves ahead.[23]

As another response to the crisis, scientists re-established long-dormant professional societies. The societies, whether new or dating to the Soviet era, lacked the cohesion and the political savvy that their Western counterparts had. After the Russian revolution, hundreds of societies and associations of specialists representing all branches and disciplines formed as scientists rushed to fill the void left by the disappearance of the tsarist system. The Geographical Society, the Russian Society of Physicists (split off from the Russian Physical Chemical Society in 1908), and other societies with pre-revolutionary roots struggled to achieve independence while having to rely on the government for funding for their publications and for approval of their charters. Between 1918 and 1925, several hundred new organizations were formed. Most were short-lived, and many were connected with local or regional history, but the most important ones were national in scope and aspirations.[24] Most of them received no money from the Commissariat of Enlightenment and had to rely on membership dues no less in this early period than in the present. These societies sought to present their interests before government and citizens alike, to protect their autonomy, and to seek greater sources of support.

The Russian Association of Physicists (RAF), founded in 1918 under Abram Ioffe, expanded rapidly in the 1920s, holding roughly biannual meetings that were increasingly international in attendance. The programs of the meetings show that Russian physicists had re-established the vitality of their discipline after the turmoil of the war and revolution, and that physicists were an integral part of international science. In 1928, for example, such leading European physicists as P. A. M. Dirac and Paul Ehrenfest joined Ioffe and scores of Soviet physicists at the fifth congress of the RAF. As part of the conference activities, they took a steamship down the Volga River, discussing new developments in quantum physics, and stopping along the way in provincial towns to address public gatherings about recent advances. Even with public engagement, several Party officials interpreted as the steamboat trip as symbolic of the divorce of the physicists from the needs of socialist construction. During Stalin's Great Break, the RAF, like other national societies, was forced to be accountable to state industrialization programs and subjugated to Party organs because of the fear of bureaucrats that experts might form some kind of locus of power or technocratic impulse independent of their control. The RAF disappeared as a unit of the Commissariat of Heavy Industry in 1931.[25]

In 1988, Sergei Kapitsa (son of the Nobel laureate Petr), a physicist who was head of a sub-department (kafedra) of the Department of Physics at the Moscow Institute of Physics and Technology, but best known for his long-running TV program *Ochevidnoe-neveroiatnoe* (meaning Obvious Yet Incredible), tried to reestablish the RAF as the Society of Soviet Physicists. He built on his experience as an active member—with Zhores Alferov and a few others—of the European Physical Society in the late 1970s and the 1980s. But once again the absence of funding prevented the reborn "RAF" from meeting more than a dozen times, although it did put out a newsletter for a while. And like any other scientific organization, it would crumble under the uncertainty of the breakup of the Soviet Union and the collapse of the economy under Yeltsin. A similar Ukrainian Physical Society experienced a rapid rebirth and an equally rapid disappearance in the 1990s. Alferov kept his distance from the Russian Association of Physicists, not because he rejected its aspirations, but because his research in nanotechnology and his activities in the Academy's leadership, in the Parliament, and at the LFTI consumed his time.

Though scientists established dozens of new professional organizations in the glow of perestroika, these organizations lost membership and steam as quickly as they had gained them, owing to the downward-spiraling economy. The societies lacked historical experience on which to build— and they lacked membership dues or other finances. Scientists resorted to protest in hopes of gaining government attention and financing. They protested at their institutes and in front of government buildings, and they submitted a series of petitions signed by hundreds of scholars. In the most extreme case, the physicist and academician V. Nechia of the Nuclear Research Center in Cheliabinsk committed suicide to protest his institute's precarious finances.[26]

Vladimir Strakhov, director of the Institute of Terrestrial Physics, formed a "Revival of Russian Science" movement in 2001. At its first meeting, he warned the attendees of the imminent collapse of Russian science, citing the 60 percent decline in researchers from 2 million (in the USSR) to 800,000 in 2000, the aging of machinery, equipment, and instruments in empty shells of institutes, and the aging of the scientific establishment that had left Russia on "the bottom rung of states with the least scientific potential—alongside Hungary, Spain, Poland, and New Zealand." Strakhov blamed lack of funding, especially low pay, since the real wages of workers had dropped 80 percent since the breakup of the USSR, so that the average spending per researcher was one-twenty-fifth that in the other industrialized states. He also reported threats to academic freedom, with the government seeking to control foreign contacts in the name of counter-espionage endeavors.[27]

The latter threat was clear in the intent of a directive, endorsed by President Putin and promoted by the FSB, that ordered scientists "to avoid any harm to the Russian state in the sphere of economic and scientific cooperation." This meant that they had to inform their laboratory and institute directors and the foreign departments of their institutes of any foreign contacts or agreements including financial support from abroad. This frightening interference into academic freedom may have seemed on the surface to be a reporting requirement, but could and almost certainly would be interpreted as an effort to restrict foreign contacts of all sorts by requiring permission ahead of time.[28]

Though the storm of controversy surrounding that order passed, it remains unclear whether or when a Russian scientist might face charges of espionage or theft of state secrets for working openly and honestly with foreign colleagues. Such cases have occurred, although it must be mentioned that the administrations of Ronald Reagan and George W. Bush also attempted to establish control over research and researchers by restricting or declaring secret what bureaucrats deemed to be sensitive material in federally funded but not classified research. American scientists vehemently protested the effort at censorship. They pointed out that slowing the dissemination of cutting-edge research (that without direct military application) would hurt the American economy and American science.

Nobel Prizes and Soviet Insecurity

On top of the pain of working in the cold shell of a building without the funds to do research, Russian scientists felt increasingly insecure and bitter about their lack of respect abroad, a situation that would persist until 2000. Every year, another raft of American Nobel laureates. Every year, Russian scientists slighted by the international community. This was the message of stories in the Russian press during the Yeltsin years. Given that Semenov, Tamm, Cherenkov, Kapitsa, and Landau received their prizes in the 1950s, the 1960s, and the 1970s for work done in the 1930s, many people concluded that geopolitics—the isolation of the USSR, its own periodic pursuit of autarky as one reason, the distaste for its foreign policies, and the lack of democratic institutions at home—were central factors in the apparent underrepresentation of Soviet scholars. These factors were reflected in underrepresentation of Soviet scientists not only in Nobel Prizes, but also in various citation indices and other measures of scientific performance. Beginning in the 1930s, Soviet leaders and scientists had introduced a series of prizes and awards in the natural sciences, the social sciences, and the humanities, in part because of international isolation.

Domestic policies also contributed to the annual Nobel lament. Aleksandra Kollontai, the Soviet Union's ambassador to Sweden, attempted to promote recognition of her country's citizens for consideration of Nobel Prizes in the 1930s. Kollontai is perhaps better known as a longtime Bolshevik who worked with Lenin from the early 1900s on, and who broke with him over the increasingly bureaucratized state in the early 1920s as a co-leader of the so-called Workers' Opposition. Kollontai's radical views about the need to liberate women from the repressive institutions of bourgeois society such as marriage continue to shock many observers of Soviet history and Marxist thought. Prevented by Stalin's conservative turn to a pro-natalist policy and resurrection of the family as an institution of social stability from advocating sexual freedom, Kollontai served her advanced

years in the diplomatic corps in a kind of exile from state politics.[29] This
helped save her from the purges that ensnared so many other old Bolsheviks. From Stockholm, Kollontai sent cables to the foreign ministry in Moscow urging Soviet leaders to nominate Soviet citizens for Nobel Prizes. But
the Soviet authorities would not officially participate in the Nobel nomination process. Several individuals of Russian descent were nominated by
non-Soviets for the physics prize between 1934 and 1944 (the biophysicist
Alexander Gurvich and George Gamow among them), in chemistry (Vladimir Ipatieff three times), and several individuals in physiology and medicine. Several Soviet scholars also nominated foreign physicists during that
period—for example, Anton Valter, a leading specialist in nuclear physics
at the Kharkiv Physical Technical Institute, nominated James Chadwick in
1936 and Enrico Fermi in 1938.[30]

At the end of World War II, the Swedish mission in Moscow turned to
the All-Union Society for International Cultural Ties with a request for Soviet nominations for the Nobel Prize. The society passed the letter along to
the presidium of the Academy of Sciences. Swedish officials wrote a number of Soviet scientists and leading cultural officials, for example the faculty
of Leningrad State University, asking for nominations in various categories
of the prize. But by the time the All-Union Society forwarded a list of potential nominators to the Central Committee, Cold War tensions led the
Party to close the USSR's borders to scientific travel and raise the flags of
ideological danger. By mid 1947, English-language scientific journals in the
USSR had ceased publication, ties with foreign societies had been severed,
censors became more vigilant, and the USSR ignored the Nobel committee.

Still, in the years 1945–1947 scholars of Russian origin, some of whom
were still Soviet citizens, gained nominations from outside the USSR for
Nobel Prizes in physiology and medicine, chemistry, and physics. Among
the physicists nominated were Gamow, Kapitsa, Vladimir Veksler (for his
work leading to the development of the synchrotron), and Dmitrii Skobeltsyn (for his work on cosmic rays). In literature, Nikolai Berdiaev, Boris
Pasternak, and Mikhail Sholokov all gained nominations. Aleksandra Kollontai was nominated for the peace prize in 1946.[31] In the period 1948–
1953, Veksler, Kapitsa, and Semenov were again nominated, as was the
biologist Nikolai Timofeev-Resovskii, who survived both Hitler and Stalin
conducting research on radiation genetics.[32]

Over the years, and especially from the Stalin era onward, Soviet officials decided to develop prestigious Soviet prizes. One postwar proposal was for an international Lenin Prize, with a huge monetary award
of 300,000–500,000 rubles, to be awarded on Lenin's birthday. Officials

asked the Academy of Sciences to advance candidates for this prize by January 1, 1948. Sometime in 1948, the authorities decided to offer Gorky and Mendeleev Prizes instead. Then in 1949, on Stalin's seventieth birthday, they offered instead a Stalin Prize ("for the strengthening of peace among peoples").[33] Among the winners of Stalin Prizes in the 1950s were the French physicist Frederic Joliot-Curie, the American actor Paul Robeson, the British biologist John Bernal, the Soviet journalist and author Ilya Ehrenburg, the German playwright Bertolt Brecht, and the Indian physicist Chandrasekhara Raman.[34]

After Stalin's death, Soviet officials and scholars engaged the Nobel Prize process much more actively. In November of 1955, at a regularly scheduled meeting of Academy's division of the physical and mathematical sciences, scientists addressed the question of the nomination of Soviet scholars for the Nobel Prize, but concluded that the prize was "not international" and decided not to make any nominations at that time.[35] Earlier that year, physicists under the leadership of Vitalii Ginzburg had celebrated the fiftieth anniversary of relativity theory. That celebration was a clear sign of their philosophical, if not their political, autonomy, since it indicated that they would no longer endure misguided ideological attacks on Einstein in any form. At the end of January 1956, at a meeting of the Politburo, Party officials suddenly passed a resolution endorsing nominations of Soviet individuals for the Nobel Prize they had received from several sources, including from the Ministry of Higher Education, the chairman of the committee of the International Stalin Peace Prize, Dmitri Skobeltsyn, and the writer Mikhail Sholokov. The resolution came too late for action. Still, in November of 1956, N. N. Semenov became the first Soviet Nobel laureate, an award warmly received by Soviet society from officialdom and scientists down.[36] According to a recent in-depth history of the Nobel Prize in the Soviet era, Semenov's prize "did not signify for the Academy complete freedom of action in the nomination campaigns; however, requisite exchange of letters with the Central Committee for each candidate advanced had ended." This small step was "one of the marks of the general liberalization of the regime." In 1956, Igor Kurchatov asked leading physicists to nominate Pavel Cherenkov, Igor Tamm, and Ilya Frank for their work on the discovery and theoretical explanation of the Vavilov-Cherenkov effect; they received the prize in 1958.

The experience of Boris Pasternak indicated that the USSR remained—both in science and certainly in arts and literature—an authoritarian regime. Pasternak was nominated seven times for the Nobel Prize in literature (1946–1950, 1957, and 1958), at first by the British Slavicist Cecil Boyer.

After the death of Stalin, Pasternak agreed to allow *Doctor Zhivago* to be delivered secretly to a publisher in Milan. The novel created a storm of interest and became an overnight best-seller. Its vivid and unflattering treatment of the Russian Revolution (with its human costs and its destruction of the intellectual class) and its riveting discussion of the evils of the Bolsheviks' and the Whites' tactics during the civil war led Party officials to condemn Pasternak and the award. The Central Committee referred to the novel's "slanderous" characterization of the Revolution, and *Literaturnaia Gazeta* followed with a negative review that signaled a national campaign of slander. Protesters (with official approval, of course) demanded that Pasternak be stripped of his citizenship, and eventually he was expelled from the Union of Soviet Writers. Naturally, he was denied permission to travel to Stockholm to receive his prize. When he died, *Literaturnaia Gazeta* published only a small four-line announcement, befitting a Party bureaucrat, not one of the most important novelists and poets of the twentieth century.[37]

Before he received a Nobel Prize for physics in 2003, Vitalii Ginzburg asked, in the title of an article, "Why don't Soviet scientists receive the Nobel Prizes they deserve?" He implied that several of the laureates in physics and chemistry from other nations may not have deserved their prizes, but he didn't suggest that the Nobel committee had intentionally slighted Soviet scientists. Rather, scores of nominations and dozens of deserving scholars made the selection difficult. Still, Ginzburg acknowledged, antipathy toward the USSR may have contributed to the underrepresentation of Soviet scientists among Nobel laureates.

Over the years, Soviet journalists and scientists wrote scores of articles addressing the apparent slight to Soviet science. During the 1990s, when Americans garnered two or three prizes in chemistry, physics, and other fields each year, the nationalistic press in Russia complained that the Nobel committee had unfairly ignored scientists (such as Alferov) who merited their recognition. Russian unhappiness over the lack of Nobel recognition dates to the first decades of the USSR, when European governments treated Bolsheviks as pariahs, as a result of which such deserving individuals as the chemist Vladimir Ipatieff were not recognized.

These critics of the actions of the Nobel Prize committee made a strong case that the Moscow physicists G. S. Landsberg and L. I. Mandelshtam, deserved a Nobel Prize. Working with Landsberg, Mandelshtam discovered and correctly interpreted the phenomenon of combinational light scattering in quartz crystal at the same time as C. V. Raman (from India) obtained similar results. Raman received the Nobel Prize, and the effect is named after him, but Landsberg and Mandelshtam produced more precise data and did so earlier. (Granted, Raman published his results in *Nature* some months before Landsberg and Mandelshtam.) Was this again some kind of slight to the USSR? Documents seem to indicate that Landsberg and Mandelshtam simply weren't nominated by their own colleagues. Ginzburg concluded that Soviet physicists were responsible for the oversight.[38]

Turning to the postwar years, Ginzburg noted that American physicists received more than 70 prizes up to the year 2000, while Soviet physicists were awarded only seven. Did this mean that Soviet physicists were ten times weaker than American ones? Ginzburg contended that structural and ideological obstacles to conducting research certainly made Soviet science somewhat weaker, especially because of the burden of focusing on applications to benefit the military, the insufficient financing of fundamental research (especially for the construction of such devices as particle accelerators and telescopes), and restrictions on publication and foreign travel related to secrecy (which in many cases left Soviet scientists unable to prove priority). Stalin-era xenophobia and autarky also prevented healthy participation in the Nobel process and even contact with the Nobel committee. Yet Ginzburg believed that Soviet scientists were seriously slighted only on two occasions: Petr Lebedev's demonstration of light pressure (1910) and A. A. Fridmann's work on relativistic cosmology (1922). And Ginzburg reminded his readers that "in Russia, it is necessary to live a long time" to achieve success. Kapitsa received the Nobel Prize in 1978 when he was 84 years old, for work completed 40 years earlier,[39] Ginzburg received his in 2003 at the age of 87, and Alferov received his in 2000 at the age of 70.

Alferov was internationally renowned by 2000, having received a number of international prizes. In 1989, he garnered the A. P. Karpinskii Prize, named after a former president of the Soviet Academy and awarded annually by the Hamburg-based FFC Fund (founded in 1931 to help scholars of eastern, central, and southwest Europe) to a Soviet scholar for outstanding achievements in the natural or social sciences, and from 1974 on in fundamental and applied research. (For a list of Alferov's prizes, see table 6.1. For a list of Soviet and Russian Nobel laureates in physics, see table 6.2.)

One of the compelling human aspects of modern science is the motivation to be first in discovery and then to confirm priority. After each Nobel award in the sciences, several scientists claim to have made the same contribution and to feel slighted. In some cases scientists have spent thousands of dollars to purchase advertisements in newspapers in Europe and the United States to make their case for priority. Often a claim of priority comes down to a few weeks or even a few days. Zhores Alferov will

Table 6.1
Alferov's international prizes.

Ballantyne Medal of the Franklin Institute (US), 1971
Lenin Prize (USSR), 1972
Hewlett-Packard Europhysics Prize for Outstanding Achievement in Condensed Matter Physics, 1978
State Prize (USSR), 1984
GaAs Symposium Award and H. Welker Medal, 1987
Karpinskii Prize (Federal Republic of Germany), 1989
Ioffe Prize (Russian Academy of Sciences), 1996
Nicholas Holonyak Jr. Award (Optical Society of America), 2000
Nobel Prize for Physics, 2000
Kyoto Prize for Lifetime Achievements in Advanced Technology, 2001
Gold Medal of SPIE, 2002
Global Energy Prize, 2005

Table 6.2
Soviet Nobel laureates in physics.

I. E. Tamm, 1958
I. Frank, 1958
P. A. Cherenkov, 1958
L. D. Landau, 1962
N. G. Basov, 1964
A. M. Prokhorov, 1964
P. L. Kapitsa, 1978
Zh. I. Alferov, 2000
V. L. Ginzburg, 2003
A. Abrikosov, 2003

excitedly discuss who he beat to the discovery of heterojunctions and how he can demonstrate that he did so.

On October 10, 2000, Alferov received a telephone call informing him that he would share a Nobel Prize with Herbert Kroemer and Jack Kilby. Beyond the immediate honor for himself and his country, for Alferov the Nobel Prize presented an opportunity to chastise the Russian government for its failure to support science and education and to protect the legacy of Soviet science. In the Parliament, Alferov criticized the Putin administration harshly. He noted that the Kremlin's proposed budget for 2001 included far more spending on free housing for parliamentary deputies—about $40 million—than on research buildings and equipment for the nation's scientists. "Just think again," he said. "Think one more time. How can it be that the draft budget foresees 50 percent higher allocation for the Finance Ministry, tax services and other financial bodies representing pure bureaucrats than for the whole of Russian science?" He continued: "How can it happen—and I hope that the deputies will support me—that the draft budget foresees allocating 1.1 billion rubles for the construction of a special block of flats for the deputies? The amount is more than four times higher than

Zhores Alferov shaking the hand of King Karl XVI Gustav of Sweden upon receipt of his Nobel Prize, December 10, 2000.

Zhores Alferov (far right) with fellow laureates Jack Kilby and Herbert Kroemer. Alferov shared the prize with Kroemer for co-discovery of the heterojunction; Kilby
was recognized for his contribution to the development of the integrated circuit.

all capital investments for Russian science. This block of flats alone might
give us an opportunity to build several laboratories." He saw this as an insult to the nation's great scientific tradition. In a telephone interview, he
told the *New York Times*: "To our pity, frequently our leadership in the old
time—and especially right now—did not understand really how important science is for Russia. I hope the awarding of the Nobel Prize will help
to change that."[40] Alferov has continued to speak out on the importance
of protecting basic science, developing a modern electronics industry, and
expanding and reforming higher education.

Alferov's attendance at the 2001 Nobel awards banquet led him to think
about whether American science was so very far ahead of Russian science
judging by the number of Nobel Prizes and other criteria. At the banquet
the attendees observed a ritual where the ambassadors of each country

Alferov with Vladimir Putin, president of Russia, in 2000.

gathered past laureates for lunch. There were a hundred Americans, but for Russia there was only Alferov. The last living Russian laureate at that time, Aleksandr Prokhorov, who shared the prize in 1964 with Nikolai Basov for discovery of the laser, was ill and died shortly thereafter. "But independent of that," Alferov said, "when our science and institutes were on the edge of death in 1992, our foreign colleagues were the first to come to our assistance. They understood the greatness of Russian science and understood its importance perhaps more than our own government."[41]

For all his criticism of the Putin administration, Alferov remained on good terms with President Putin until recently, and he occupies an honored place in the country. He is much more visible than either of the other two recent Russian Nobel laureates in physics, Vitalii Ginzburg and Alexei Abrikosov (the latter a specialist in condensed matter physics who currently works at Argonne National Laboratory). In 2005, on the occasion of his 75th birthday, Alferov received a Presidential Award for "meritorious service of the nation" in support of science and for his active participation in the legislative branch of government.[42]

Rebuilding Russian Science and Education: Alferov's New Center

In the day-to-day practice of higher education, as well as in the Parliament, Alferov has shown an untiring commitment to reestablishing the
level of excellence that Russian science had achieved in the Soviet era, and
to finding better ways to organize the training of young specialists. Toward
this end he has used his contacts in the scientific community and political circles, and he has drawn on his own funds from various international
prizes, including the Nobel. The physical symbol of Alferov's commitment
to science and new forms of higher education is his Scientific Educational
Center (NOTs).

The huge facility occupies roughly five acres of land. It is less than a ten-
minute walk from the LFTI and the Polytechnical Institute. The towering,
open, modern architecture signifies aspirations to the scientific heavens
and reflects Alferov's view that knowledge is universal, omniscient, and
powerful.

Working with architects, Alferov designed two magnificent pink faux
marble buildings, each topped by a cathedral-like roof with a massive col-
umn in the center dressed to look like a pencil; light standards around the
building also employ the pencil motif, with mercury vapor lights on top.
The first building, with an area of nearly 150,000 square feet, opened in
September of 1999. It houses classrooms, computer center, research facilities, a conference center with two auditoriums (one seating 100 and the
other 390), and a sports facility with a swimming pool. Alferov wanted
students to understand the central place of athletics and exercise in higher
education. The second building, dedicated in 2005, holds more classrooms
and two research laboratories, one for semiconductor heterostructures and
one for photoelectric and solar energy research.[43] Alferov has met with architects concerning final drawings for a third building, but now wonders
when it will be built. "My ability to convince has aged," he told me.[44]

Alferov established NOTs to integrate science and education in the areas
of physics and information technology. The center reflects his vision of the
uncompleted work of Peter the Great, who intended to bring science and
education to Russia under one roof. NOTs consists of a lyceum for grades
8–10 and the Academy Physical Technical University, both of which are
connected to the LFTI formally and informally. The lyceum was founded in
1987 toward the end of creating the proper institution to identify and train
promising young scientists with a classical education. Leaving no doubt
about his dedication to the reform of higher education in Russia, Alferov
used one-third of his Nobel Prize money ($75,000) to establish a scholar-
ship fund for students at the lyceum. With additional support from the

Siemens Corporation, all students attend the lyceum without charge. At present, the roughly 180 lyceum students take classes six days a week. At the lyceum and the university, students have a heavy load: of six to eight hours of physics instruction a week, eight to ten hours of mathematics, four to six hours of English, and also another foreign language. Weekly laboratory work at the LFTI, the Sechenov Institute of Evolutionary Physiology, the Medical Institute of Cytology and elsewhere supplements this study.

Between 1989 and 2007, nearly a thousand students graduated from the lyceum, of whom roughly one-half entered a university or an institute associated with the LFTI and one-third entered a university not associated with the LFTI. They selected as majors not only physics but also biology, medicine, geology, and even the social sciences and humanities, and some have continued their studies at universities in the United States, Germany, Israel, Sweden, England, France, and Denmark.[45] Among the 40 faculty members are a total of four doctors and candidates of science, eight individuals who have won awards from the International Soros Science Education Program, and three Honored Teachers of Russia. V. V. Fedorov teaches neutron physics, Dmitrii Varshalovich teaches astrophysics, and Edward Tropp (academic secretary of the Petersburg Scientific Center) often lectures on history and philosophy of science.

A highlight of academic life for the students and faculty of the lyceum is an annual lecture series that has included lectures by such luminaries of science and culture as Boris Zakharchenia, Sergei Kapitsa, K. G. Skriabin, Evgenii Velikhov, Edward Tropp, and of course Alferov himself, and not only on scientific topics. On February 10, 2006, Alferov spoke about "Stalingrad: The Turning Point in the Great Fatherland War." The lyceum staff also convene an annual Sakharov Scientific Symposium for the students, for which Alferov serves as chairman. Roughly 200 school children present papers.[46]

The Academy Physical Technical University was another of Alferov's dreams, and a source of great frustration. Alferov presented this project to President Putin in 2001, and Putin promised his support for an annual budget of roughly $120,000, but the project languished for years in the Ministry of Finance and the Ministry of Justice. The university opened on NOTs territory in 2002, but not until 2006 did the bureaucrats approve the final charter for it, and its future remained clouded by budgetary and other uncertainties.[47] Alferov blames bureaucrats for their glacially slow comprehension of the fact that human capital is much more important than natural resources to Russia's future. Like the lyceum, the Academy University has departments that reflect the strengths of its founder and his colleagues: the physics and technology of nanostructures (under Alferov), astrophysics (under Dmitrii Varshalovich), and neutron physics (under

V. A. Nazarenko). Alferov proudly notes that this institution is the first university within the Academy of Sciences.[48]

Alferov has consistently called for tying research and education together in something like a Western university. According to Graham and Dezhina, the USSR "enthroned the idea of the non-teaching research institute more than any other leading scientific country in the world." Further, it separated teaching, fundamental research, and applied research in different networks—universities, the Academy of Sciences, and branch institutes. It "tightly combined" highest honors (full membership in the Academy) and the actual administration of scientific research. This put it out of step with world trends, according to more and more scientists.[49]

Alferov has urged Russia's businessmen to follow his example of philanthropy and to "think about the future of Russia." That future, Alferov believes, is "not for the Berezovskiis and Gusinskiis [wealthy oligarchs] but for the Alferovs and their students." He has asked businessmen to contribute to the Alferov Fund for students of the lyceum and students of other Petersburg educational institutions, the basic goal of which was "the support of national education and the Russian school of fundamental research in the area of physics which is vitally necessary for the development of the national economy."[50]

In trying to encourage philanthropy on the part of businessmen, Alferov has generally been unsuccessful, as have other individuals who have sought to resurrect private-sector grant giving. First, the nascent tradition of philanthropic organizations disappeared with the Russian Revolution nearly a century ago. Few wealthy entrepreneurs in tsarist Russia sought to follow the example of the Rockefellers, the Harrimans, Andrew Carnegie, or Henry Ford. The Lendentsov Fund for the Advancement of Positive Science operated from 1907 until World War I and supported Nikolai Umov, Ivan Pavlov, and other individual scientists, as well as laboratories and journals. But the Ledentsov Fund disappeared when the Bolsheviks confiscated virtually all capital in the name of the worker. Second, Russia's present-day businessmen seem psychologically predisposed to accumulate vast sums of money without any thought of philanthropy. Finally, Russian tax law does not offer any tax breaks to not-for-profit foundations—something that makes little sense to bureaucrats.

Service to Russia and the Putin Administration

Alferov's service to Russia goes beyond his efforts to establish dynamic new structures for higher education, to expand the research foundation for nanotechnology, to lobby the government for adequate funding, and to cure bureaucrats of their myopic insistence on seeing in science only for its short-term contributions to economic growth. Indeed, officials in the Putin administration, like many in the administration of George W. Bush, saw science and technology only for their ability to harness natural resources for short-term economic gains. What did the Putin presidency meant for Alferov in terms of the funding and health of educational and research programs and the extent of his committee service in the Parliament?

In his last years as president, as a symbol of the failed effort to establish effective rule, Boris Yeltsin frequently changed prime ministers. On the eve of his resignation, Yeltsin appointed Vladimir Putin, an obscure former KGB official from Leningrad, prime minister. Putin was a disciplined, non-drinking, non-smoking, no-nonsense intelligence officer who had worked mostly in East Germany. He had a reputation of being decent and honest, but he was largely unknown. He had left the KGB in 1990 and had become an ally of Anatoly Sobchak, the liberal mayor of St. Petersburg, who was his tutor at Leningrad State University. In 1996, Prime Minister Anatoly Chubais recommended Putin for a job in the Yeltsin administration, in which he rose to the position of deputy chief of staff. In July of 1998, he was appointed the head of the FSB (the successor to the KGB). Yeltsin appointed him prime minister in August of 1999, then resigned five months later on December 31. Putin became acting president and ordered elections to be held on March 26, 2000, in which he received roughly 53 percent of the vote. Before this time he had no experience in elected office.[51]

Although they knew little about him, many people embraced Putin as Russia's savior. In contrast with the impetuous and unpredictable Yeltsin, whose government had become increasingly stained by the air of corruption, Putin would put the economy in order, bring stability to the political system, and restore Russia's pride. In his two terms as president, Putin accomplished precisely this, yet also chipped away at several democratic reforms. He established control by fiat over the regional governors, who had become, in his mind, too independent. He gained the power to appoint them to avoid the creation of fiefdoms in the periphery. About 90 percent of all media outlets are now state controlled. In the name of rapid development of Russia's great natural resources, Putin disbanded the federal environmental protection agency and gave the provincial governments responsibility for enforcement of weakened laws but without adequate staff or personnel, and the FSB closely monitors—some would say interferes with—the operations of non-government organizations, especially those with political agendas or strong foreign contacts.[52]

As part of the effort to rebuild Russia's sense of itself as a superpower, Putin supports increased expenditures on space exploration, nuclear power, and the military, while the infrastructure of provincial cities has collapsed from lack of funding. Many provincial municipal governments have often been unable to provide basic services, let alone snow removal, on the basis of extremely modest local tax revenues. Since 2006 the federal government has recognized the need to make significantly greater resources available in the provinces and municipalities, although Moscow continues to be a kind of "black hole" of power and money. Although Putin has developed a sound tax policy and has reformed banks, many Russians, fearful of a repeat of the 1998 economic collapse in which they lost all of their savings, keep their savings at home. And in the international arena, whereas at first Putin welcomed US efforts to fight terrorism, even permitting US military aircraft to fly over Russian territory, and used the idea of fighting terrorism to justify harsh policies in Chechnya, later he adopted a much more combative tone regarding Russia's interests as the administration of George W. Bush badly misplayed Russian sympathies over the 9/11 attacks. Putin strongly rejected NATO expansion and criticized the Bush administration's effort to build a missile-defense system in Poland as destabilizing and not in Russia's interests. (The vast majority of the world's physicists and engineers, Alferov among them, recognize that the costly system—more than $60 billion has been spent on developing an anti-ballistic-missile system—simply will not work and is a disgusting waste of resources.) Putin rejected the criticisms of Secretary of State Condoleezza Rice about the backsliding of democracy in Russia as ill-informed, and ridiculed European and American commentary about lack of transparency in Russian economic institutions—given the crisis that enveloped Wall Street beginning in 2007.

Yet the close connections between oil magnates, state-run businesses for the exploitation of oil, gas, and minerals, and the Putin presidency led many observers to suggest that Russia was now run by an oligarchy that resembled tsarist and Soviet political culture in many ways. A small number of men had centralized power in their hands. The secret police cooperated intimately with the government. The state was the critical actor, since citizens lacked the experience to develop political institutions to participate in civic culture, and the majority of people seemed apathetic.[53]

Early in the Putin administration, Alferov hoped for good working relations with the president. He agreed to chair a commission of the Parliament on the handling of nuclear waste, but the stuff in question turned out to be nuclear waste with a twist. In 2000 and 2001, after relatively limited parliamentary debate, the Russian government passed a law to permit the importation of spent nuclear fuel from other countries in exchange for cash. The nuclear ministry, Rosatom, estimated receipts from foreign governments at some $30 billion. The government promoted the bill with the promise that Rosatom would use the income to engage in extensive remediation of radioactive waste problems in the nuclear enterprise dating to the early Soviet period.[54] Officials have made great progress in clean up, nuclear materials inventory and control, and so on since 1991, in part through joint and multi-lateral programs with the European Union, Japan, and the United States.[55] Yet Russian lawmakers explicitly excluded the public from debates over the efficacy and safety of importing spent fuel rods. Tens of thousand of Russians signed petitions asking for a referendum on the law before its enactment—more than enough, under Russian law, to require a referendum. But the courts rejected the petitions with the claim that not all the signatures could be verified. These events indicated the government's determination to build scores of new reactors by 2030, perhaps 40 in Russia and the rest ordered by other countries, with or without public support. Construction is underway also on the first of a dozen or so "floating" nuclear power stations. Nuclear power also will enable Russia to export more oil and gas to Europe.[56]

Perhaps to ensure public support for the fuel-rods-for-money program and deflect criticism of it, and perhaps out of determination to rejuvenate the nuclear industry, Putin appointed Alferov the head of a parliamentary committee on Issues Concerning the Import into the Territory of the Russian Federation of Spent Nuclear Fuel. Alferov's involvement in the committee reminds us to recognize the generally strong support for nuclear power among Russia's scientists and engineers.[57] The committee included twenty members, five each as representatives of the president, the Council of the Russian Federation, the Parliament, and the government, with Alferov as chairman.[58] Alferov's public authority as a Nobel laureate, the pride with which citizens speak about his achievements, and his defense of science made him a logical choice. On his appointment, Alferov proclaimed "The creation of this commission is not to calm fears, but rather to move in the direction of society to review the issues surrounding the reprocessing of spent nuclear fuel." The committee had responsibility for overseeing nuclear imports, although it relied on the self-interested Rosatom for information about the imports. Alferov saw the earnings as possibly providing the government with funds to build up the nation's decaying scientific infrastructure. Alferov did not know how the moneys earned by the import of nuclear fuel would be applied, yet hoped they might go to solar power research.[59]

Alferov and other committee members had a very limited mandate. They did not address the general state of nuclear safety in Russia. Before the State Nuclear Inspectorate (Gosatomnadzor) was weakened under Putin, its head, Iurii Vishnevskii, had criticized the transport and disposal of spent fuel from nuclear submarines, suggesting that it had left many regions of the nation "on the edge of ecological catastrophe." The Ministry of Defense, not Gosatomnadzor, has responsibility for the radiation safety of military sites, but Vishnevskii asserted that "in reality" there was a "complete absence of effective inspection for nuclear and radiation safety."[60] On top of dozens of waste facilities, millions of gallons of liquid and millions of tons of solid nuclear waste, and 50,000 tons of spent nuclear fuel remain as the legacy of Soviet nuclear power within Russia's borders.

In the summer of 2006, Alferov asked to step down from his position as head of the committee on spent nuclear fuel, feeling that he had fulfilled his parliamentary duties and desiring to focus on NOTs. Putin named as his replacement a vice president of the Russian Academy, Nikolai Laverov. Alferov had already asked several times to be freed from committee service because of his commitment to other issues. The Russian "Greens" believed that Alferov was removed because he had already fulfilled his role as a "buffer" between the government and society. Laverov, a geologist and a specialist in nuclear issues, had no doubts about the place of nuclear power in Russia's energy future.[61] He supported a Russian plan to reprocess the spent fuel to extract uranium and plutonium for use in a mixed-oxide fuel to be burned in a new generation of liquid-metal fast reactors.[62] He approved of the importation of spent fuel and the establishment of an "international center for the delivery of fresh nuclear fuel to participating nations as well as procedures for recovering the spent fuel to prevent it from falling into the wrong hands."[63] Alferov now seems to be content with Rosatom pursuing an aggressive new construction program, although perhaps not to the extent that the pro-nuclear lobby desires.

The Frustrations of Parliamentary Inaction

Alferov increasingly criticized the government's machinations in the Parliament as a continued assault on science. In April of 2002 he resigned his post as chairman of the subcommittee on science in protest. He considered further service on it "useless." He described a "campaign of anti-Communist hysteria which had unfolded recently in the Parliament" as "the first characteristic line of fascism." He noted that on April 3 the Parliament had removed members of the Communist fraction from seven of nine committees on which they served, leaving Communists on only the committees on culture and tourism. They were offered posts on committees for non-government organizations and religious organizations but refused them. Alferov asserted that the removal of Communists from these committees led to the replacement of qualified individuals with other individuals who lacked a similar level of experience or "professionalism."[64]

Until this time, Alferov had reveled in the opportunity to serve in the Parliament. He did not find it to be so much a drain on his time, and he saw it as another opportunity to serve his country. Even in his mid seventies, he managed to keep up a very busy schedule. However, he told me that in the 1980s, when he was much younger and was serving in the Congress of People's Deputies, the extensive travel was easier to bear. He spent two or three days in Moscow each week, and the Congress met only for ten sessions a few times per year. He coordinated his attendance at meetings at the Academy's presidium with those in the Congress. When in Leningrad, he spent roughly two hours each day at the Scientific Center and the rest at the LFTI. During the mid 1990s, when he was elected to the Parliament as a deputy representing the centrist Our Russian Home Party, he continued with his heavy schedule. He left St. Petersburg on Monday evenings, arriving by train in Moscow on Tuesday morning. He spent Tuesday morning at the Academy's presidium, Tuesday afternoon in the parliamentary committee on education and science, and Wednesday again at the Parliament. Thursday morning he returned to Petersburg and the Academy. He frequently spent not only Friday but also the weekend at the institute.[65]

The excitement of perestroika engaged intellectuals of many stripes to consider the promise of Soviet socialism in the twenty-first century, not its demise. Some individuals, Alferov among them, anticipated a kind of New Economic Policy, with market mechanisms playing a role but with the state maintaining its control over the commanding heights of the economy and ensuring proper levels of funding for health care, education, and science. He did not anticipate confronting the constant effort of fighting to preserve the Academy. What he calls ineptitude and an uncaring attitude toward the Soviet legacy of science and education pushed him into the Communist fraction.

In the 1990s, when Our Russian Home fell apart, Alferov was well on the way to returning to the Communist roots of his family. Alferov's growing frustration with the Putin administration reflected a different intellectual atmosphere than during perestroika and different economic and political challenges than those of the 1980s. Only the Communist fraction, Alferov believed, would carry on the necessary struggle to protect the nation's scientific heritage. "Each day the government is thinking about something else for the Academy to give up. They say, 'We'll take the hospitals, the

publishing house, the House of Scholars, the Hotel, the university'—my
university—because 'that's not science.'" Alferov criticized the Putin administration for failing to understand infrastructure as crucial to science.
For them, "science" was the building of a research institute itself, nothing beyond its walls. In the Academy's vast real estate holdings they saw
equally vast, if temporary profits through fire sale or real estate development. Alferov believes this is not only short-sighted but would be another
heavy blow to science. For example, without the hotels, how would people
attending conferences, graduate students, and visiting scholars pay for
their accommodations at market rates in Russia today? A night in a cheaper
hotel in St. Petersburg or Moscow cost at least $150, more than the average
Russian scholar could afford. If the Nauka publishing house were confiscated, Alferov continued, how could scientists be certain of competent,
timely publication of technical results? Alferov believes that the Putin administration has no sense of the value of supporting basic science for the
future of the economy. Only colleagues abroad understand the dilemma:
"We are appreciated only by foreign science."[66]

At a meeting of the presidium of the St. Petersburg Scientific Center
in 2004, Alferov again warned that, because of low salaries and lack of
financial funding for research, the grave crisis of Russian science had deep-
ened. More and more scholars left science for business or positions abroad.
Even more dangerous than brain drain was the resulting generation gap.
An entire layer of the most productive scientists from 35 to 50 years old
had essentially disappeared. Alferov said that within 25 years "not one scientific researcher will remain" unless government policies changed. Only
scientists with international reputation maintained the necessary level of
support, often only by virtue of their administrative posts. At the same
time, Academy institutes had been forced to rent out space within their
walls to commercial businesses to earn money for research and salaries.
In 2003 roughly 116,500 individuals worked in Academy institutes; a year
later the number had fallen slightly to 115,400, but, according to Alferov,
the most productive had remained. Alferov acknowledged that the budget
for science had increased recently, but only because the entire government
budget grew. Remember that into the twenty-first century, funding for scientific research as a share of GDP was perhaps one-sixth its level on the eve
of the breakup of the USSR.[67]

Alferov feels he must engage in an annual "rite of autumn" to criticize
Putin administration policies for the sciences. In autumn 2004, the Education and Science Ministry developed a plan ("the national Concept of
participation in Managing State Scientific and Research Organizations") to

privatize several R&D institutes. This would have reduced the number of state institutions from roughly 2,000 to between 100 and 200 in 2008, and enabled the government to cut state funding for science by half. Many institutes already had essentially no funding, while others stumbled along on contract funding. Alferov saw this plan as more evidence that the ministry acted in a "tactless and arbitrary manner" without even consulting with scientists on the matter. He declared that to follow the plan meant "Russian science will soon be dead and buried." Eduard Tropp, scientific secretary of the St. Petersburg Center, likened the tactics to Soviet collectivization,[68] apparently meaning the effort to consolidate without thought to human costs.

The Kremlin's effort to control the Russian Academy of Sciences continues. For many observers, the need to reform the Academy from the outside must proceed since the Academy's leadership has been unable to engage meaningful reforms from within. Those critics believe that Academy institutes had become too large and too conservative, and many of its institutes no longer performed at a world level. Downsizing and competitive funding were the best ways to identify leading programs. They argue that in the twenty-first century having an Academy of Science that combines research and honorific functions is anachronistic and fails to bring new blood and approaches to the laboratory. They find Tropp's criticism overstated, to say the least. Academy personnel, from their side, see most attempts by the government to change the Academy as misguided and geared to control of the Academy's wealth, not as sincere efforts to improve performance.

In September of 2006, Minister of Science Andrei Fursenko, who began his career at the LFTI in the late 1980s as one of Alferov's deputies, announced an amendment to the science law that would give President Putin the right to approve future Academy presidents and changes to its charter. According to *Science*, "While some officials say that the specific endorsement of President Putin means that RAS presidents will have greater authority to lead the organization, others believe this new law is an attempt to enable the government to acquire the Academy's assets and turn the organization into an honorific society such as the US Academy of Sciences." Perhaps this signals the end of the battle to shrink the Academy "to a level commensurate with the quality of its researchers and make the bloated organization more cost-effective." While the Academy will continue to receive roughly one-third of its budget, which amounted in 2006 to $1.27 billion, from the government, it will also have to be more accountable to the state, including paying property taxes, whereas American educational institutions are tax exempt but often voluntarily pay property taxes. Since many institutes could not pay the tax, the government has paid it for them. All of this adds up to a bleak outlook. According to the former science minister Boris Saltykov, "It is not a secret that the Academy is dying." No longer the site of world-class research, it is an ossified, dusty old dinosaur.[69]

Less-Than-Nobel Politics

On top of the struggles in the Academy, Alferov faced struggles within the Leningrad Physical Technical Institute. Alferov, who had given up the post of director of the LFTI in 2003, was asked to give up his position as the institute's director for science in the spring of 2006. Formally, the academic council had merely fulfilled an Academy resolution of 2005 that prohibited a person from holding that position who did not occupy the post of director. And, at the time of his departure from the institute ruling structure, Alferov was, after all, older than Ioffe when Vavilov informed him of the need to "retire." The geriatric structure of the ruling elite in the Academy has been a concern of many reformers. But, according to insiders in the institute, Alferov's departure was the result of an effort to end "dual power" that had existed since the appointment of a new director, Andrei Zabrodskii, in 2003. ("Dual power" has an interesting connotation in Russian politics. In the spring of 1917, the provisional government that assumed power after the fall of the tsarist regime found itself in the ultimately fatal position of having to share power with the workers' and peasant's soviets that were springing up, especially the Petrograd Soviet.)

In the case of the physicists, dual power existed as soon as Alferov quit the post of director of the LFTI and named Zabrodskii as his successor, but kept his other position. Making things more complicated, the Academy presidium approved the reorganization of the LFTI with the institute becoming the "St. Petersburg Physical Technical Scientific Educational Center of the Russian Academy of Sciences"—still under the chairmanship of Alferov. The center consisted of the Scientific Technological Center of Microelectronics, the Lyceum, and the Academy Physical Technical University of the RAN.[70] Hence you had a director of a center (Alferov) and a director of an institute (Zabrodskii), each with a different view of the future of the LFTI in whatever new form it took. Not surprisingly, they differed over the disposition of financial, manpower, space, and other resources. Several LFTI physicists worried that Alferov's command of political authority and resources at the LFTI, at NOTs, and the Academy presidium might have a negative effect on their desire to expand the LFTI's programs in new directions. In one case, NOTs had received $2.5 million for new apparatus, but the Academy's leadership counted it as part of the LFTI's stock,[71] although essentially without benefit to the LFTI. Alferov makes it clear that he gains nothing financially from his activities, but gave and continues to give substantial portions of his salary and prize moneys "to the development of science."[72]

The LFTI physicists also worried that Alferov had designs on some 3,000 square meters of laboratory space, on equipment, and on employees. After all, Alferov had been a powerful scientific administrator for 25 years and was used to getting his way. While willing to part with several employees since the responsibility for salaries would no longer be theirs, LFTI physicists worried about the disposition of material resources that were essential to the institute's future.[73] When Zabrodskii expressed concern over institute resources, Alferov pointed to the common interests of the LFTI and NOTs and reminded him that each was a financially independent legal entity.[74] Unfortunately, the interested parties discussed these differences heatedly in local and national newspapers—for example, in *Poisk*, a weekly devoted to the scientific community. They then tried to smooth over differences as minor disputes that occur in scientific "families" from time to time.[75] But feelings may stay be hurt. (Because of illness, Alferov did not attend the ninetieth-anniversary celebration of the LFTI, held in October of 2008.)

In fact, such a dispute among colleagues and friends reflects not real personal animosity but the depth of the financial crisis facing Russian science and the uncertainties how government "reforms" will play out. As the second-largest academic institute in Russia, the LFTI has 71 laboratories, 130,000 square meters of buildings and other facilities, 220 doctors of science, and 600 candidates of science. Its staff has shrunk from 2,039 to 1,813 employees, and will certainly shrink more as finances for salaries and research remain extremely tight. The LFTI leadership claims to need roughly 3 billion rubles (approximately $100 million) to reestablish the institute's research facilities and programs, but the government will not and perhaps cannot support any institute at that level. In 2007 the institute received the equivalent of only a million dollars for modernization and a million dollars for repairs.[76]

Another source of concern is the fact that the LFTI has aged; the average age of the researchers is approaching 60. The directors of the LFTI understand the need to attract promising young scientists to fill new positions. Alferov and Zabrodskii therefore are on the same page here, even when they do not so say in public. NOTs will be an important conduit of young talent for the entire complex of lyceum, institutes, laboratories, and university.[77]

Minister of Science Andrei Fursenko attended the next meeting of the academic council after Alferov ceased being director of science for the LFTI. Fursenko's presence at the meeting was interesting on two counts. Fursenko apparently asked the scientists to pay more attention to innovation projects. This pressure for applications is a major concern of Alferov today. He believes it reflects the diminution of the value of fundamental science by the government.[78] For his part, Fursenko considers the Soviet scientific heritage "too large." He believes the Russian government cannot and should not support R&D at a level anywhere near that of the Soviet era. Rather than see the paucity of Russian Nobel laureates as an insult, he suggests it reflects an objective evaluation of the quality of science. He believes that Soviet science was "a 'safety valve' for people of free mind" rather than a sacred enterprise. Science has shrunk to its "natural level" of meeting the economic demands, and from Fursenko's point of view the government's scientific and technological policy should be based a view of science as a branch of the economy.[79]

A second aspect of this story reveals that in post-Soviet Russia, just as in the Stalin, Khrushchev, and Brezhnev eras, personal relationships between all-powerful institute directors and government bureaucrats are crucial to the conduct of science. In the late 1980s, Alferov appointed Iurii Kovalchuk and Andrei Fursenko, two young scientists and activists in the Komsomol, as deputy directors of the LFTI to help plan the institute's financial path during perestroika and its pressures for research institutes to become self-financing. These men later became important "facilitators" of the Putin era—to put it mildly. Kovalchuk got a degree in physics from Leningrad University in 1974 and, after completing his doctorate in physics, joined the LFTI. As one of Alferov's deputy directors from 1987 to 1991, he advanced the idea of creating a series of innovation firms within the institute to produce and sell technologies. Alferov did not support this plan. The two young specialists either were asked to go or left on their own volition, seeing only dead ends, though apparently Alferov would have permitted Fursenko to stay on. Kovalchuk then moved through a series of different high-tech businesses before entering the banking industry; he is now one of the directors of the very large and well-connected Rossiia Bank, which is connected with the highly profitable Severstal Enterprise[80] (one of a number of strategic metal facilities around which Russia's oligarchs have tempered their wealth and power).

The upward trajectory of Kovalchuk's career path probably was linked to his close association with Vladimir Putin, who before becoming prime minister was active in the St. Petersburg administration. In 1995 Kovalchuk joined the St. Petersburg Municipal Intergovernmental Commission on Enterprises with Foreign Investments under Putin's chairmanship, and in 1996 he was one of eight founders of a cooperative dacha venture on Komosomolsk Lake in the Priozersk region of the Leningrad region, along with Putin and Fursenko.[81] Apparently, Kovalchuk proposed dividing the LFTI into applied, physics, and technical sectors, with himself heading the applied sector—an artificial relationship that Alferov could not endorse because it would create new barriers between basic and applied research.[82]

Putin's (and now Medvedev's) Minister of Higher Education and Science, Andrei Fursenko, generally has a solid reputation among scholars, even if they disapprove of the administration's science policy overall and recognize him as a bureaucrat. Fursenko, the son of a well-known historian and academician, hoped to enter the mathematical mechanical department of Leningrad University, but according to one source he did not get in and chose the less prestigious mechanics department. According to classmates, he was a solid but pedantic student who also showed little interest in social activities or sports. In his junior year Fursenko asked to join a practicum at the LFTI. He did well in it, eventually earning a spot as a junior scientific worker in the institute's department of computer hydrodynamics and kinetics. By all accounts he was thrilled to be there, and he joined a Party committee that coincidentally was under the chairmanship of Alferov. The same source indicates that Fursenko convinced the administration to do more to acquire computers for the institute. The department in any event became the hydrodynamic laboratory, and Fursenko its director at the level of "senior scientific worker." Alferov was impressed by his energy and business acumen, and brought Fursenko and Kovalchuk into the director's office to develop strategies for the LFTI to generate more income in the increasingly rocky times of perestroika. This led to their proposal to create an "innovation belt" based on institute research, transform a number of institute positions into contractual positions, and create an institute bank. They warned that there would be dire financial consequences if the institute did not listen to them. Given what we know about the disastrous first years of the Yeltsin administration, it is safe to safe that dire finances would have resulted in any event. "Our new ideas," Fursenko diplomatically recalled, "were in conflict with the traditions of academic science. Maybe we were a bit unnecessarily radical. But there were two ways to solve the situation. Either we began to struggle within the institute with people whom we considered to be our teachers and old comrades, or we simply leave. And we decided to leave. Believe us, this was a very difficult decision."[83] Most of the tension connected with his departure is long forgotten among the principals, but important political and scientific disputes over resources, power, and authority persist.

Having left the institute, the young scholars determined to show that they were correct and attempted to profit from their own scientific knowledge. Toward that end they created a special innovation center. But to make money they needed start up money, and they did not even have an office. Fursenko, strangely, thought it crass to make money through commercial sales of any sort (what kind of center might it be without profits?), and therefore decided to turn to the town fathers, who approved the creation of a regional fund for scientific-technology cooperation of St. Petersburg. The fund was intended to secure financing for scientific projects that had future commercial promise. Fursenko showed skillful handling of people and ideas in a very difficult financial situation. His managerial skills caught the attention of Petersburg officials.[84]

In the fall of 2001, Vladimir Putin appointed Fursenko deputy minister for industry, science, and technology. Fursenko recalled that initially he did not wish to transfer to Moscow, but he found the work with Minister Ilya Klebanov engaging, and the work precisely what he wished to do since President Putin was committed to science policy that supported innovation. Of course, in the view of such academicians as Alferov, this science policy underplayed the importance of basic research and threatened the entire edifice of the Academy of Sciences. Officials added funds to the annual federal budget "for innovation," and the ministry established a special department for innovation policy. In 2003 Klebanov became the plenipotentiary of the Northwest Federal Region, and Fursenko became Minister of Science.[85]

Fursenko immediately pursued new federal policies toward science that seemed to threaten the Academy. Fursenko has stated on a number of occasions that the government was unlikely to increase resources for science. Rather, "science was never perceived as part of the market in the past. We need to change this attitude and make it more marketable. Our scientific institutions from Soviet times are important resources, but you all need to be aware, that the government only has a finite amount of resources and that is not going to change."[86] In a few years, Alferov, Fursenko, and others would debate not only appropriate federal policy for basic science but also innovation policy in the field of nanotechnology. Family disputes have thus has been played out both in Petersburg and at the highest levels of the Russian government. However, these disputes indicated not only different views of the role of science in contemporary Russia but also revealed the increasing vertical integration of politics and society under Putin, including a new view of the place of scientists in institutions under federal control.

Alferov and other scientists have been fighting with the Putin administration for years over the government's spoken intention of changing the charter of the Academy of Sciences. Government spokesmen state that they wish to increase the efficiency of scientific research, while scientists believe that the government wishes to turn the Academy into a bureaucracy geared to making money. Vitalii Ginzburg, who disagrees strongly with Alferov about the Soviet legacy, shares his criticism of the attempt to install government bureaucrats as "business managers" to profit from science. "The managers," Ginzburg said, "will begin to define politics in science, and the Academy will be doing errands for them."[87] Ginzburg suspected the issues were also tied to personality conflicts. Officials simply believed that the Academy generally was not capable of managing resources efficiently, while scientists argued that the government wished to appoint bureaucrats whose lack of competence in scientific matters would put their enterprise at great risk.

In some ways, Fursenko's comments on the appropriate relationship between science and government suggest similar—and ongoing—debates between scientists and officials over autonomy of researchers, accountability to the government for funding, and the value of basic science in a variety of settings—in the United States, Japan, the European Union, and elsewhere. What is fascinating in this case is how those debates resemble those that took place in the Stalin era. Under Stalin, the government centralized science policy in major bureaucracies. It moved administration from the Commissariat of Enlightenment to the Commissariat of Heavy Industry, leaving no doubt about the state's interest in, and pressure for, applied science. The government criticized scientists for engaging in "ivory tower reasoning" rather than research that might be of immediate benefit to the proletariat. It forced greater political and ideological accountability on the Soviet Academy of Sciences, transferring its seat to Moscow, physically and psychological closer to the Party apparatus. Although forced to pay homage to applications and to subjugate their research in many ways to the glories of "planning," scientists found increasing largesse for their programs in the huge budgets given to the industrialization program, and the Academy maintained its authority and a modicum of autonomy. The ongoing debates about the place of science in modern Russia remind Alferov what was lost when the Soviet Union disappeared: the authority of basic research and money, and he believes that current officials have less understanding of, and appreciation for basic science than did Soviet bureaucrats. Yet even if the Putin and Medvedev governments have dangerously treated science as just another branch of the economy, and basic science has lost its prestige, officials do not pursue autarky or impose strict ideological claims on research paradigms.

Nanotechnologies for Russia's Future

Fursenko and Alferov share a vision, although they differ on the path to its realization. Among others in science and government, Alferov wants the nation to embark on an extensive—and expensive—program to support the development of various nanotechnologies. Nanotechnology is the science of assembling devices out of individual atoms or molecules. It involves the application of physical laws to control of matter on the atomic and molecular scale. Nanotechnology research will advance fundamental understanding of materials at the subatomic, atomic, and molecular levels. Scientists expect a broad range of devices and materials with unprecedented speed, efficiency, and exactness. Nanotechnologies will include high-performance materials, more efficient manufacturing processes, increased computer storage capacity, energy-saving and energy-producing devices and insulators, and biomedical applications ranging from efficient drug-delivery systems to cancer therapies, biosensors, and skin. Russian scientists and leaders see nanotechnology as an investment in the future and as a way to demonstrate that Russia's greatest capital is its minds, not its oil and gas.

Alferov laments the failure of the government to support microelectronics, including nanotechnologies, at the proper level. For him they are the "engine of progress." But the notion of a national nanotechnology program raises a series of questions, including questions about who will finance it and at what level, how to distribute funds and to whom, how to involve the private sector (for example through grants or contracts), and who will provide the scientific leadership. And Russian officials, businesspeople, and scientists are not the only people debating whether and how to establish a federal program to support a specific area of industry. This debate goes on in the United States and other countries too. In the mid 1990s, the administration of President Bill Clinton pursued a nanotechnology initiative, with federal funds. Yet the record indicates that it has been difficult to predict "winners" in any new field of technology.[88]

Fursenko supported the idea of creating a kind of "Russian Silicon Valley," and has pushed it among leading politicians, including members of the Council for Science and High Technology that advises the president on science and technology policy.[89] In theory, former President Putin endorsed the idea of creating an integral national innovations system with a developed infrastructure, a technology market, and legal protection of the results of intellectual work. The plan grew out of a March 2002 joint meeting of the President's Council for Science and High Technology, the Presidium of the Council of the Federation, and the Security Council. The participants adopted a program that included more words than results: "On the Political Standpoint of the Russian Federation on the Issue of the Development of Science and Technology until 2010, and Its Long-Term Perspectives." These "long-term perspectives" indicated the government's determination to reform the Academy into an institution accountable to pressure to conduct research with rapid economic returns, with three-quarters of the government's budget for fundamental research and applied science shifted to "technology of governmental significance."[90]

In 2004 Fursenko announced that the government was considering launching a Federal Target Program for nanotechnology, perhaps in response to persistent criticism in the Parliament by Alferov and others of the low level of funding for science and technology; the funding would come from the government's Research and Development Program, itself with a miserly budget of $49 million in 2005.[91] In December of 2006, Prime Minister Mikhail Fradkov seemed to announce such a Federal Target Program aimed at nanotechnologies in the amount of $1.1 billion through 2010, and the creation of a federal agency, Rosnanotekh, to pursue applications. But at a press conference staged to discuss this award, no minister seemed to be able to say what a nanotechnology was. Whether the government would fund this program, or, if funded, it would fuel economic growth, was uncertain.

Russian efforts in nanotechnology had a long history within Academy institutes, even if organizational reforms to accelerate R&D were relatively recent. In the spring of 2008, at a meeting of the presidium of the RAN, Zhores Alferov optimistically discussed research and organizational developments in nanotechnology. He noted that the Academy had established a commission on nanotechnology in July of 2007, and that the commission had begun its work in October. The commission's members touted a broad range of potential applications in materials science, energy, electronics, biology, diagnostics, and education. The commission had six sections that roughly corresponded to these six areas. Researchers at institutes from St. Petersburg to Moscow to Siberia and the Far East were involved, and thus far had submitted hundreds of proposals for funding totaling approximately 34 billion rubles ($100 million). Alferov stated that NOTs had been involved in nanotechnology for more than eight years and had attracted other research institutes and universities to the endeavor. He predicted that a nanotechnology market exceeding a trillion rubles (more than $30 billion) would appear within ten years. He pointed to the growing role of the Russian government in nanotechnology through Rosnanotekh, a government corporation that would enable Russia to keep pace with similar public-private efforts in the United States.[92]

When the Russian government eventually approved funding, sometime in 2007, it anointed Mikhail Kovalchuk, the director of the Kurchatov Institute of Atomic Energy, as the scientific director of nanotechnology industrial R&D. To many observers, this was a surprising choice. Kovalchuk, a specialist in materials science, was not a full member of the RAN, and therefore was prohibited from being appointed an Academy vice president, an appointment legally required for official designation as nanotechnology tsar. Instead, his title was "acting" vice president of the Academy for nanotechnology. Further, because Putin been involved in machinations surrounding the Academy, rumors circulated that Kovalchuk might become the Academy's next president.[93] Kovalchuk's uncertain position between microchips and money may therefore have mirrored the power struggle between the Kremlin and the RAN.

Making the situation more curious, Mikhail Kovalchuk may have been drawn into the inner circle of Kremlin politics by his brother, Iurii, who, it will be recalled, had worked with Fursenko at the LFTI before becoming president of Rossiia Bank, which had connections to the massive and lucrative Severstal metals operation. The close ties have led a number of Russian scientists to fear a lack of transparency in nanotechnology funding. Five Russian companies have received support of their investment projects and are already producing devices based on nanotechnology. One of those companies is Severstal.[94]

For Alferov, the lengthy process that led to the formation of Rosnanotekh serves as a reminder of missed opportunities. He points out that since the collapse of the USSR the Russian microelectronics industry has fallen further behind its counterparts in Europe, Japan, and the United States in productivity, efficiency, and size. In 2003, a German firm, Zander, proposed building a microchip factory in Russia at terms favorable to the government that would have enabled it to close this gap and would have benefited the scientific community greatly. Alferov points out that the government's reluctance to put funds into science and technology has also resulted in declines in Russia's aviation industry and other leading industries relative to their international competitors. He also points out that even in the oil and gas industry—the major source of federal revenues and perhaps of Russian influence—the productivity of labor and technology has fallen far behind that of leading American and European firms. Furthermore, while Western firms have moved far ahead in diodes, lasers, solar cells, and other technologies based on Alferov's research, the Russian microelectronics industry stagnated.[95]

In 2005, in another incident that left Alferov vulnerable both in the court of public opinion and in the opinion of government officials, a jury that he had chaired awarded him the Global Energy Prize.[96] His friends Nick Holonyak, without doubt a deserving awardee, had received the prize in 2003. When the trustees of the prize announced new winners in April of 2006, they voted to strip Alferov of his chairmanship of Global Energy's foreign committee. Alferov blamed his removal in part on Sergei Yastrzhembskii, President Putin's press secretary, and on his resistance to interference in the prize by its major state corporate sponsors, including Gazprom and Surfutneftegaz. Alferov says he was surprised to hear that he had won the prize in 2005, and that the reason he didn't refuse it was that he feared offending his colleagues. Not surprisingly, his acceptance offended some people. "Everyone on the Board knew that granting the award to Alferov meant its collapse," said one person connected with the board of trustees.[97] Several scientists worry that the conflicts over the LFTI and the Global Energy Prize have weakened Alferov's authority as vice president of the Academy of Sciences and chairman of the St. Petersburg Scientific Center.[98]

The threats to Russian science, the failure of the government to treat scholars with respect, the attitude of bureaucrats toward science as one part of a larger economy that ought to serve industry, and the perceived lack of support from his institute and his country led Alferov firmly into the Communist orbit. Yet his membership in the Communist fraction seems somehow a distant tie, perhaps because none of the leading members of the fraction ever speaks about science. Articles published in *Pravda* during the Twelfth Extraordinary Conference of the Communist Party of the Russian Federation on September 22 through September 24, 2007, had not one word on science or education. The Communists advanced a list of 523 candidates for election to the Parliament in December of 2007, headed by the chairman of the Central Committee of the Party, Gennadi Ziuganov, the chairman of the agribusiness union, N. M. Kharitonov, and Zhores Alferov. But there were no words about or by Alferov, who attended the conference. Zuganov referred to need to defend the interests of workers, to fight for pensioners who lived on the edge of poverty and starvation, to understand the needs of the small businessman while fighting against the oligarchy, and to reject Putin's United Russia Party as the party of "oil socialism." But, again, nothing about science and education.[99] Alferov remains in the Duma as deputy representing the Communist fraction in 2008. He will continue to travel to Moscow, to NOTs, and to the LNTs. However, he finds his obligations more and more draining.

Whither the Laureate?

From the Scientific Educational Center, Alferov doggedly continues to pursue both cutting-edge science and a more dynamic institutional form for science and higher education. He believes that nanotechnology holds the greatest promise of all applications in solid state physics. He remains head of the laboratory of nanoheterostructure physics at the LFTI. He has sponsored a series of meetings to gather international scholars and heighten the interest of his government in supporting the development of nanotechnology. In June of 2002, for example, the institute organized a conference with seventeen sessions on nanostructure technology, including sessions on wide-bandgap nanostructures, microcavity and photonic crystals, silicon-based nanostructures, quantum wells and superlattices, and quantum wires and dots. Physicists (including the Nobel laureate Leo Esaki, with whom Alferov launched this symposium in 1993) presented more than 200 papers. Alferov gave a paper (co-authored with N. A. Maleev and physicists from the Technical University of Berlin) on "MBE Growth of Low-Threshold Long-wavelength QD Lasers on GaAs Substrates" that considered recent results on molecular-beam epitaxy growth of the quantum dot InGaAs/GaAs heterostructures for long wavelength lasers on GaAs substrates.[100]

At the fifth session of the Duma, called to order at noon on December 24, 2007, Alferov, as the senior deputy, had the honor of delivering the opening address. He pointed out the additional pleasure of knowing that the main expert on the question of his age, his wife, continued to consider him a young man.[101] Turning to the important issues facing Russia, Alferov urged the deputies not to ignore Russia's scientific heritage and the problems that scientists could solve. "The Duma," he noted, "begins its work in a very complex period for Russian and world history. My colleagues in scientific and world society recently have been discussing above all else such vitally important problems as the problem of steady state development of our planet, global warming, and the creation of clean contemporary energy technologies." Earlier in the year, he had attended a special meeting, held in Potsdam, of Nobel laureates who agreed that the most significant problem facing humanity was steady state development of the economy and

Zhores Alferov and Tamara Georgievna (née Darsksaia) on their wedding day, 1967.

in particular the energy economy. He pointed out that the major source of this problem was that the wealthiest nations consumed far more than their share of resources.[102]

Russia, Alferov suggested, mirrored the situation elsewhere in the world, with billionaires dominating consumption and pensioners and others scraping to make ends meet. Alferov urged the deputies to "begin the battle with poverty with a change of tax laws from a flat tax to progressive scheme" and to increase per capita spending to the level of that in the advanced developed countries. He argued that Russia could not continue to rely on resource exploitation as its main strategy for economic develop-ment and ensure steady rational economic growth. "The only road to a solution of this basic economic problem," he insisted, "is the creation and development of an economy based on knowledge, it is the development of an economy based on the achievements of science and technologies richly endowed with science." He called on his fellow deputies to fund research in nanotechnology immediately and effectively. He reminded those who thought that funding for R&D made little economic sense that science and technology generated discoveries that generated income and then generated tax revenues. He concluded with an appeal to the deputies

Zhores Alferov with his son, Vanya (Ivan Zhoresovich), 1972.

Zhores Alferov with his parents, 1981.

Zhores and Tamara Alferov on a beach at Varadero, Cuba, in 2006. He was in Cuba to receive an honorary degree in physics that had been awarded in 1987.

to remember that "their authority depends on the authority of the laws that they consider and pass which is determined by the quality of those laws." And those laws had to represent the needs and interests of the Russian people.[103]

Zhores Ivanovich Alferov's father participated in the Great War, the Russian Revolution, and the Russian civil war. Ivan Karpovich Alferov and his wife, Anna Vladimirovna, endured the pressures of rapid industrialization and collectivization, and often ate quite poorly because of harvests that failed owing to government policies, while millions of peasants perished of starvation. Their two sons, Marx and Zhores, saw the Great Terror envelop the nation in the 1930s, and the entire family wept when Marx was killed in one of the last great tank battles of World War II. About 20 million people died during the war, and more than 1,500 towns were reduced to rubble. Stalin required the survivors to turn immediately to rebuilding the country. For a number of years, many citizens lived in hovels. Another famine broke out in 1946. Large numbers of heroic veterans who had had contact with the West were arrested and sent to work in labor camps.

The Alferovs with Fidel Castro, president of Cuba, in 2006.

Scientists—especially those interested in genetics, relativity theory, and quantum mechanics—were subjected to hostile scrutiny by unknowing ideologues, and were hounded to conform to epistemological and scientific norms. Many were arrested, and scores of talented scientists lost their lives in prison camps. Zhores Alferov was fortunate to come of scientific age after the worst challenges to physicists had passed. Yet he witnessed the purge of the founder of the Leningrad Physical Technical Institute, Abram Ioffe, and he saw Andrei Sakharov and Vitalii Ginzburg, two fellow Nobel laureates, suffer personally and professionally for their struggles against an authoritarian regime.

For many people, the human costs of industrialization paled in comparison with the glorious achievements of the "hero projects" of Stalinism—the Moscow Metro, the Baltic White Sea Canal, the steel city Magnitogorsk, the massive hydroelectric power stations built along the Volga. Zhores Alferov himself was thrilled to see the Tsimliansk reservoir and power station. He continues to believe that only the Soviet system could have produced the Kuibyshev hydroelectric power station, the Obninsk reactor, Sputnik, and his own heterojunctions. And he laments the destruction by Presidents Yeltsin and Putin of what he considers to be the greatest of the Soviet Union's achievements: its scientific and educational system. Still, he has continued to serve Russia in the Parliament, in scientific administration, and in informal advisory capacities.

As would have been the case in another society, the discovery for which Alferov received his Nobel Prize had many sources: excellent training among a cohort of talented and driven students; superb, far-sighted scientific leadership; collegiality and democratic relations within the LFTI and his laboratory; adequate government support; and a mindset that refused to accept limits to knowledge. Discovery will occur even when authorities interfere with academic freedom. Yet Alferov wants his scientific legacy to be—rather than heterojunctions and the communications, solar, and other technologies that have arisen from them—his university, his scientific-education center, and his lyceum. Just as he had an opportunity to study with leading scholars doing cutting-edge research, he wishes to give young people in Russia today an opportunity to contribute their own ideas to science, to be infused with the kind of enthusiasm that he had, and to push the development of nanoheterostructures and solar cells to greater efficiencies.

Afterword and Acknowledgments

In some ways this book dates back to a lecture I attended in Cambridge, Massachusetts, in 1980. Sheldon Glashow, co-recipient of a 1979 Nobel Prize for work on quantum gravitation, gave a talk on the background to his theory and mentioned in passing the work of a relatively unknown Russian relativist, Alexander Friedmann. In 1922, Friedmann had proposed equations that suggested that the universe could be expanding or contracting as a function of the amount of matter in it. The equations were a challenge to Albert Einstein, whose initial theory of general relativity (1916) had included a cosmological constant that posited a static universe. Einstein seems to have had no reason for adding the constant except for the difficulty of apprehending an expanding universe. Friedmann (a geophysicist, meteorologist, and mathematician) suggested a non-static universe as a mathematical possibility, not a requirement.

After his lecture, I approached Professor Glashow to ask more about Friedmann. He said that he really did not know much about the man. Finding little information about Friedmann in American libraries, I wrote to Petr Kapitsa, the 1978 Nobel laureate in physics, to ask his help. Kapitsa kindly passed my letter to Viktor Frenkel', a specialist at the Leningrad Physical Technical Institute and the leading historian of Soviet physics.

Frenkel' had written extensively about Friedmann, the history of his institute, and the history of Soviet nuclear physics. He worked tirelessly in archives to illuminate achievements by Soviet scholars that were shielded from full examination owing to the secrecy of the state. Frenkel' wrote back to me sometime in 1980, telling me what he knew about Friedmann and suggesting that I work in the Soviet archives, although acknowledging that it would be difficult to arrange such work. In 1984 and 1985, as a graduate student working on the history of Soviet physics, I gained an IREX (International Research and Exchanges Board) fellowship to spend nine months in the Soviet Union doing dissertation research on the Petersburg-Leningrad

physics community. I met Viktor Frenkel' during that time. I will forever
be indebted to him for serving as my advisor in the Soviet Union and for
opening his home to me.

In March of 1985 I spent seven weeks doing research in Leningrad. Viktor and his wife, the biologist Olga Cherneva, made their home near the
LFTI a second home for me. Over the years, I spent a number of evenings
with the Frenkels, and on later research trips I stayed overnight with them.
At one dinner at their apartment in the spring of 1985, I met Viktor Golant, a plasma physicist, his wife, and Zhores Alferov and his wife, Tamara
Georgievna.

Over dinner, I sensed that Alferov was a generous and open man, was
devoted to physics, nation, and family, and was a wonderful storyteller.
We sat around a gorgeously set table in the Frenkels' dining room, eating a
marvelous Central Asian plof (lamb rice pilaf), drinking dry Georgian wine,
and exchanging stories and anecdotes.

I have known Zhores Alferov for over 20 years, have met with him at
his home and his various offices, in Moscow, and even in the United States.
In 2002 or 2003, I determined to write a biography of him, and in it to
discuss the history of Soviet physics and Soviet science generally. This has
been a difficult task on several counts. First, Zhores has read every chapter,
and we have disagreed on some points of interpretation and emphasis,
including Zhores's criticism of my "anti-Soviet" attitude in places. Also,
I wanted to include details of events that Zhores believes have no place
in his biography. Some of those events reflect the tension between such a
big personality as Zhores, the politics of research, the challenges of raising
funds to do so, and the occasional disputes that arise in the process, and
so I have included them. Throughout the entire process of research and
writing, Zhores has been remained open, engaging, and helpful, and we
have grown closer and more informal. I am deeply indebted to Zhores and
his family for the access they have given me to their personal histories,
photographs, and papers, and to Zhores in particular for the time he has
provided to me.

Second, as is often the case with scientists at the pinnacle of success,
Zhores Alferov has engendered the jealousy of a number of former colleagues, acquaintances, and friends, some of whom are no longer on speaking terms with him. Others find his return to Communism distasteful;
many of them suffered personally or lost family or friends in the Stalinist
purges. As I hope I have made clear in this book, Alferov's politics derive
in large part from his devotion to science and higher education and his
belief that Russia's future depends on proper financial support of science.

He believes that the Putin and Medvedev administrations have failed to provide scientists with either the funding or the autonomy they ought to have.

Others have criticized Alferov for ignoring their contributions to the research that led to his Nobel Prize, and for disputes that have arisen between Alferov and his colleagues at the LFTI over how the institute should proceed in a time of financial uncertainty. Yet all of them have been honest with me about their deep respect for Alferov.

I am thankful to Academic Secretary Andrei Shergin and Director Andrei Zabrodskii for giving me open access to the LFTI's archives and library, and to the professional staff of the archives and the library. They have been forthcoming in their comments and suggestions for this biography. I am grateful especially to Eduard Abramovich Tropp, Scientific Secretary of the Petersburg Scientific Center, a physicist, and a historian of science, for bureaucratic assistance and for critical comments on the substance of this book. Arseny Berezin of St. Petersburg, a physicist now retired from the LFTI, shared reminiscences and insights and offered critical comments on chapters 2, 3, and 5. Herb Kroemer read chapter 4 in an early form.

Mort Panish twice provided extensive critical comments on chapter 4. Bruce Parrott helped me frame the issues in chapter 2. Michael Gordon and Loren Graham each read a chapter and suggested improvements, and Michael then read the entire manuscript, offering comments as an "anonymous reviewer." Spencer Weart provided keen insights. Stephen Fortescue, whose work on Soviet science policy I have long admired, made a number of pertinent suggestions on chapter 5. Alix Hui's perusal of chapter 6 helped me improve the narrative and the analysis immeasurably; I may impose on her to read my next manuscript in its entirety. Svante Lindquist, director of the Nobel Museum, and Michael Sohlman, director of the Nobel Foundation, introduced me to the world of the Nobel Prize. Jed Buchwald patiently encouraged me to complete the manuscript.

I would like to close with a special mention of Katya Chukaeva, a talented linguist at Pomor State University, who translated Boris Slutskii's poem "Fiziki i Liriki" for chapter 3, and who died of a stroke at only 21 years of age in June of 2009.

Notes

A Brief Note on Sources

Interviews with Zhores Ivanovich Alferov over the course of four years were a major source for this biography. These interviews supplemented and clarified much of the autobiographical material that Alferov has published in a variety of journals and books, in particular in his *Nauka i Obshchestvo* (1995). Alferov and the current directors of the Leningrad Physical Technical Institute (LFTI) gave me complete and unrestricted access to Alferov's papers—scientific articles, popular-science articles, speeches, and interviews. In addition to these materials, the library of the LFTI and Alferov's own office saved clippings from local, regional, and national newspapers and journals about Alferov, his colleagues, and the LFTI, to which I was given access. Among the documents in the archives of the LFTI (cited in notes as A LFTI), I used five-year and annual plans and reports of research; correspondence between LFTI scientists and various Party, educational, scientific, branch industry, and other organizations; statistical reports; personnel folders; and additional folders of newspaper and journal clippings for the entire period of Alferov's life. —PRJ

Introduction

1. On the rise of industrial research and development in the US, see David Noble, *America by Design* (Knopf, 1977).

2. Robert Kohler, *Partners in Science: Foundations and Natural Scientists, 1900–1945* (University of Chicago Press, 1991).

3. On the impact of these changes on the physics community in the United States, see Daniel Kevles, *The Physicists* (Knopf, 1977).

4. I hesitate to list any of the superb histories of Soviet science and technology that have inspired me to write this book for fear of leaving any one of them out. They

have all been important to me. The authors of these studies include but in no way are limited to Mark Adams, Viktor Frenkel', Loren Graham, David Holloway, David Joravsky, Nikolai Krementsov, Susan Solomon, and Douglas Weiner.

5. Paul Josephson, *New Atlantis Revisited* (Princeton University Press, 1997).

6. Interview with Alferov, January 17, 2008, St. Petersburg, Russia.

7. Zhores Alferov, "The History and Future of Semiconductor Heterostructures from the Point of View of a Russian Scientist," *Physica Scripta* 68 (1996): 32–45.

Chapter 1

1. Zhores Alferov covers much of the information in this chapter in his autobiography, which I then verified and complemented in a series of interviews with him. The section on Zhores Alferov's parents is based on pp. 71–75 of Alferov, *Nauka i Obshchestvo* (Nauka, 2005).

2. Boris Mironov, "The Development of Literacy in Russia and the USSR from the Tenth to the Twentieth Century," *History of Education Quarterly* 31 (1991), no. 2: 229–252.

3. On Russian industrialization, the Trans-Siberia railroad, and Witte's role, see Stephen Marks, *Road to Power* (Cornell University Press, 1991) and Theodore von Laue, *Sergei Witte and the Industrialization of Russia* (Columbia University Press, 1963).

4. Jacob Metzer, "Railroad Development and Market Integration: The Case of Tsarist Russia," *Journal of Economic History* 34 (1974), no. 3: 529–550.

5. For an eyewitness account of the second duma, see Bernard Pares, "The Second Duma," *Slavonic Review* 2 (1923), no. 4: 36–55.

6. GAAO (State Archive of Arkhangelsk Province), F. 1863, op. 1, d. 13, l. 6 ob.

7. Among the many fine books on the Russian Revolution is Sheila Fitzpatrick's *The Russian Revolution* (Oxford University Press, 1982).

8. For reminiscences of the problems facing the government in the lead-up to elections, see Alexander Kerensky, "The Policy of the Provisional Government in 1917," *Slavonic and East European Review* 11 (1932), no. 31: 1–19.

9. GAAO, F. 1863, op. 1, d. 13, l. 7.

10. Abraham Ascher, "The Kornilov Affair," *Russian Review* 12 (1953), no. 4: 235–252.

11. Interview with Alferov, St. Petersburg, March 4, 2008.

12. On the famine in Revolutionary Russia, see Maurice Hindus, *Humanity Uprooted* (Jonathan Cape and Harrison Smith, 1929), pp. 355–357. On the American Re-

lief Administration, see Bertrand M. Patenaude, *The Big Show in Bololand* (Stanford University Press, 2002).

13. http://www.hrono.info/biograf/kaledina.html.

14. GAAO, F. 1863, op. 1, d. 13, l. 7.

15. Israel Getzler, "Lenin's Conception of Revolution as Civil War," *Slavonic and East European Review* 74 (1996), no. 3: 464–472.

16. See also Alferov, *Nauka i Obshchestvo*, pp. 77–81.

17. Robert Lewis and Richard Rowland, "Urbanization in Russia and the USSR, 1897–1966," *Annals of the Association of American Geographers* 59 (1969), no. 4: 776–796.

18. Richard Rowland, "Geographical Patterns of Jewish Population in the Pale of Settlement of Late Nineteenth Century Russia," *Jewish Social Studies* 48 (1986), no. 3/4: 207–234.

19. Yuri Slezkine, *The Jewish Century* (Princeton University Press, 2004), pp. 105–203.

20. Julian Batchinsky, Arnold Margolin, Mark Vishnitzer, and Israel Zangwill, *The Jewish Pogroms in Ukraine: Authoritative Statements on the Question of Responsibility for Recent Outbreaks against the Jews in Ukraine* (Friends of Ukraine, 1919).

21. In 1943, in *Red Star*, the correspondent and novelist Vasilii Grossman wrote: "People arriving from Kiev say that the Germans have placed a cordon of troops around the huge grave in Babi Yar where the bodies of 50,000 Jews slaughtered in Kiev at the end of September 1941 are buried. They are feverishly digging up corpses and burning them. Are they so mad as to hope thus to hide their evil traces that have been branded forever by the tears and the blood of Ukraine, branded so that it will burn brightly on the darkest night?"

22. Eric Haberer, "The German Police and Genocide in Belorussia, 1941–1944. Part II: The 'Second Sweep': Gendarmerie Killings of Jews and Gypsies on January 29, 1942," *Journal of Genocide Research* 3 (2001), no. 2: 207–218.

23. Daniel Romanovsky, "Chashniki: Essay on the Mass Murder of the Jews in Belorussia (1942)," *Vestnik Evreiskogo Universiteta v Moskve* 1 (1992): 157–199; Vyacheslav Selemenev, "Kolyshki—A 'Shtetl' in the late 1930s," *Jews in Eastern Europe* 45 (2001): 48–72; Arkadii Zeltser, "The Shtetl During the 'Great Watershed' of 1929–1931: The Case of Vitebsk region," *Jews in Eastern Europe* 46 (2001): 5–33; Zeltser "Jews in the Political Life of Vitebsk, 1917–1918," *Jews in Eastern Europe* 34 (1997): 5–27; Monica Bohm-Duchen, "The Road from Vitebsk, 1887–1907," *European Judaism* 30 (1997), no. 1: 55–63.

24. http://home.alphalink.com.au.

25. Aleksandra Shatskikh, *Vitebsk: Zhizn' Iskusstva, 1917–1922* (Iazyki Russkoi Kul'tury, 2001).

26. *Marc Chagall on Art and Culture*, ed. Benjamin Harshav (Stanford University Press, 2003); *Marc Chagall and His Times: A Documentary Narrative*, ed. Benjamin Harshav (Stanford University Press, 2004).

27. See also Alferov, *Nauka i Obshchestvo*, pp. 81–85.

28. Rene Fulop-Miller, *The Mind and Face of Bolshevism* (Putnam, 1927), p. 194. See also Alferov, *Nauka i Obshchestvo*, pp. 80–81.

29. GAAO, F. 1863, op. 1, d. 13, ll. 10–11.

30. Lars Lih, "Bolshevik Razverstka and War Communism," *Slavic Review* 45 (1986), no. 4: 673–688.

31. For a Russian biography of Jean Jaures, see Nikolai Molchanov, *Zhan Zhores* (Molodaia Gvardiia, 1986).

32. On his life and career, see M. I. Smirnov, "Admiral Kolchak," *Slavonic and East European Review* 11 (1933), no. 32: 373–387.

33. Of the many engaging studies of Stalin, his era, rapid industrialization, and collectivization, see Adam Ulam, *Stalin: The Man and His Era* (Viking, 1973).

34. On how the Stalinist cultural revolution affected research and development, see David Joravsky, "Soviet Scientists and the Great Break," *Daedalus* 83 (1960), no. 9: 562–580.

35. Interview with Alferov, St. Petersburg, March 4, 2008.

36. Loren Graham, *The Ghost of the Executed Engineer* (Harvard University Press, 1993).

37. See Matthew Payne, *Stalin's Railroad: Turksib and the Building of Socialism* (University of Pittsburgh Press, 2001).

38. Sheila Fitzpatrick, "Stalin and the Making of a New Elite, 1928–1939," *Slavic Review* 38 (1979), no. 3: 377–402.

39. James McClelland, "Proletarianizing the Student Body: The Soviet Experience During the New Economic Policy," *Past and Present* 80 (1978), August: 130–132.

40. GAAO, F. 1863, op. 1, d. 13, ll. 1–5.

41. See Lewis Siegelbaum, *Stakhanovism and the Politics of Productivity in the USSR, 1935–1941* (Cambridge University Press, 1988).

42. GAAO, F. 1863, op. 2, ed. khr. 14, ll. 1, 11, 26, 39–40, 43, 60.

43. On life in Stalingrad, see Alferov, *Nauka i Obshchestvo*, pp. 88–90.

44. On Stakhanovism and the Moscow metro, see Lazar Kaganovich, *Pobeda Metropolitena—Pobeda Sotsializma* (Transzheldorizdat, 1935) and idem., *Voprosy Zheleznodorozhnogo Transporta v Sviazi so Stakhnovskim Dvizheniem* (Partizdat, 1936).

45. Robert Conquest, *The Great Terror* (Macmillan, 1973).

46. For the move to Siberia, see Alferov, *Nauka i Obshchestvo*, pp. 90–92.

47. In the novella *Sofia Petrovna*, Lydia Chukovskaia captures the sense of loss of a family member during the Great Terror and the belief of citizens that they were innocent of crimes while all others must be guilty.

48. Alferov, *Nauka i Obshchestvo*, pp. 92–98.

49. Leonid Leonov, *Soviet River* (Dial, 1932).

50. For example, in *Russkii les* (*Russian Forest*)—published in 1953 just after Stalin's death—Leonov anticipated disputes over the environmental costs of Soviet development projects that arose in the Khrushchev and Brezhnev eras with a sophisticated discussion of the debate how best to tame the nation's great forests between the two leading characters, who are forest rangers.

51. Boris Thomson, *The Art of Compromise* (University of Toronto Press, 2001), pp. 99–124.

52. http://www.kirjasto.sci.fi/majakovs.htm

53. Source: http://www.marxists.org/subject/art/literature/mayakovsky/1929/my-soviet-passport.htm.

54. For greater detail on science policy in the early Soviet period, see Josephson, "Science Policy in the Soviet Union, 1917–1927," *Minerva* 26 (1988), no. 3: 342–369.

55. David Noble, *America by Design: Science, Technology, and the Rise of Corporate Capitalism* (Oxford University Press, 1987).

56. Kristie Macrakis, *Surviving the Swastika: Scientific Research in Nazi Germany* (Oxford University Press, 1993); *Physics and National Socialism: An Anthology of Primary Sources*, ed. Klaus Hentschel (Birkhäuser, 1993).

57. On Lunacharsky's role in early education policy, see Sheila Fitzpatrick, *The Commissariat of Enlightenment: Soviet Organization of Education and the Arts under Lunacharsky, October 1917–1921* (Cambridge University Press, 1970).

58. On Glavnauka and NTO, see Josephson, "Science Policy in the Soviet Union" and V. Kosnichenko and E. Shoshkov, "Shtab Nauki: K 70-letiiu NTO VSNkh," *Leningradskii rabochiii*, December 9, 1988, p. 10. For Ipatieff's reminiscences of working for the Bolsheviks, specifically on NTO, see V. N. Ipatieff, *The Life of a Chemist* (Stanford University Press, 1946). On Palchansky and his fate, see Graham, *Ghost of the Executed Engineer*.

59. Gorky's organization was the Free Association for the Advancement and Dissemination of the Positive Sciences (1916–1918).

60. On the intrigues and purges at the Kharkiv-based physics institute, see Iu. V. Pavlenko and Iu. N. Ranyuk, *Delo UFTI, 1935–1938* (Feniks, 1998).

61. Viktor Frenkel' and Paul Josephson, "Sovetskie fiziki—stipendiaty rokfellerov-skogo fonda," *Uspekhi fizicheskikh nauk* 160 (1990), no. 11: 103–134.

62. For an excellent biography, see Gennady Gorelik and Viktor Frenkel', *Matvei Petrovich Bronshtein* (Nauka, 1990), also published as *Matvei Petrovich Bronstein and Soviet Theoretical Physics in the Thirties* (Birkhäuser, 1994).

63. Interview with Alferov, St. Petersburg, March 4, 2008.

64. Robert McCutcheon, "The 1936–1937 Purge of Soviet Astronomers," *Slavic Review* 50 (1991), no. 1: 101–110.

Chapter 2

1. Harrison Salisbury, *The 900 Days: The Siege of Leningrad* (Da Capo, 1985).

2. On the environmental costs of the transformation of the Ural region into a center of military industry, see Josephson, "Industrial Deserts: Industry, Science and the Destruction of Nature in the Soviet Union," *Slavonic and East European Review* 85 (2007), no. 2: 294–321.

3. For the following section, see Alferov, *Nauka i Obshchestvo*, pp. 99–103.

4. http://www.mishalov.com/Molotov22June41.html

5. Radio address by Stalin, July 3, 1941, as transcribed and translated by *Soviet Russia Today*; available at http://www.mishalov.com.

6. G. A. Goncharov, "Chislennost' i razmashchenie 'trudarmeitsev' na Urale v gody velikoi otechestvennoi voiny," in *Promyshlennost' Urala*.

7. A. A. German, *Respubliki Nemtsev Povolzhiia v Sobytiiakh, fakhtakh i dokumentakh* (Gotika, 1996); Ann Sheehy, *The Crimean Tatars, Volga Germans and Meskhetians: Soviet Treatment of Some Minoritiies* (Minority Rights Group, 1980).

8. Alferov, *Nauka i Obshchestvo*, pp. 105–106.

9. Ibid., p. 108.

10. Zhores Alferov has published the letters of Marx Alferov in several places. For extracts from those letters and a discussion of Marx's service in the Red Army, see Alferov, *Nauka i Obshchestvo*, pp. 109–122.

11. Geoffrey Roberts, *Victory at Stalingrad: The Battle That Changed History* (Longman, 2002). See also Georgii Konstantinovich Zhukov, *Marshal Zhukov's Greatest Battle* (Harper and Row, 1969).

12. Interview with Zhores Alferov, St. Petersburg, October 24, 2007.

13. On Alferov's travels to Khil'ki and his efforts to chronicle Marx's military career, see Alferov, *Nauka i Obshchestvo*, pp. 126–134.

14. On the return to Belarus, see Alferov, *Nauka i Obhshchestvo*, pp. 134–152.

15. Nicholas Ganson, *The Soviet Famine of 1946–47 in Global and Historical Perspective* (Palgrave Macmillan, 2009).

16. Nikolai Krementsov, *Stalinist Science* (Princeton University Press, 1997).

17. Such corruption was a signal feature of Soviet society. See Konstantin Simis, *The USSR: The Corrupt Society* (Simon and Schuster, 1982).

18. Karl C. Berkhoff, *Harvest of Despair: Life and Death in Ukraine under Nazi Rule* (Belknap, 2004).

19. Wendy Lower, *Nazi Empire-Building and the Holocaust in Ukraine* (University of North Carolina Press, 2005).

20. Zhores Medvedev, *The Rise and Fall of T. D. Lysenko* (Doubleday, 1971); David Joravsky, *The Lysenko Affair* (Harvard University Press, 1970). See also N. N. Vorontsov, "Poslednii akt tragedii genetiki," in Vorontsov, *Nauka, Uchenye, Obshchestvo* (Nauka, 2006), originally published in *Teatr* (1988, no. 3: 59–64).

21. Interview with Alferov, St. Petersburg, October 24, 2007.

22. Salo Wittmayer Baron, *The Russian Jew under Tsars and Soviets* (Macmillan, 1964); Jonathan Brent, *Stalin's Last Crime: The Plot Against the Jewish Doctors, 1948–1953* (HarperCollins, 2003); Joshua Rubenstein and Vladimir Pavlovich Naumov, *Stalin's Secret Pogrom: The Postwar Inquisition of the Jewish Anti-Fascist Committee* (Yale University Press, 2001); Arno Lustiger, *Stalin and the Jews: The Red Book* (Enigma Books, 2003).

23. Alferov, *Nauka i Obshchestvo*, pp. 163–182.

24. Aksel Berg, *Aleksandr Stepanovich Popov (k 50-letiiu Izobreteniia Radio)* (Gosenergoizdat, 1945); *Izobretenie Radio A.S. Popovym: Sbornik Dokumentov i Materialov*, ed. Aksel Berg (Izdatel'stvo Akademii nauk SSSR, 1945).

25. G. M. Krzhizhanovskii, editor, *Stroitel' Pervykh Gidroelektrostantsii SSSR—Akademik Genrikh Osipovich Graftio, 1869–1949* (Izdatel'stvo Akademii nauk SSSR, 1953).

26. For example, *Komissiia po edinitsam mer. Sbornik Rabot*, ed. Mikhail Shatelen (Izdatel'stvo Akademii Nauk SSSR, 1938) and *Laboratornye raboty po elektromagnitnomu poliu*, second edition, ed. N. P. Glukhanov and K. I. Krylov (Academy of Sciences, 1957).

27. http://www.eltech.ru/english/general/history.html

28. For one study of the effort to "modernize" minority nationalities in the former Soviet Union, see Yuri Slezkine, *Arctic Mirrors* (Cornell University Press, 1994).

29. On the Stalinist Plan to Transform Nature, see Paul R. Josephson, *Industrialized Nature* (Island, 2002), chapter 1.

30. Alferov, *Nauka i Obshchestvo*, pp. 172–175.

31. In *How Life Writes the Book: Real Socialism and Socialist Realism in Stalin's Russia* (Cornell University Press, 1997), Thomas Luhausen discusses Azhaev's career and his writing of *Far from Moscow*.

32. Varlam Shalamov, *Kolyma Tales* (Penguin, 1994).

33. Werner Hahn, *Postwar Soviet Politics: The Fall of Zhdanov and the Defeat of Moderation, 1946–1953* (Cornell University Press, 1982).

34. Nikolai Krementsov, *The Cure: A Story of Cancer and Politics from the Annals of the Cold War* (University of Chicago Press, 2002).

35. Lenin All-Union Academy of Agricultural Science (VASKhNiL), *The Situation in Biological Science: Proceedings of VASKhNiL, July 31–August 7, 1948* (Foreign Languages Publishing House, 1948). See also T. D. Lysenko, *New Developments in the Science of Biological Species* (Foreign Languages Publishing House, 1951); I. E. Glushchenko, *Akademik T. D. Lysenko—Vydaiushchiisia Sovetskii uchenyi: Stenogramma Publichnoi Lektsii, prochitannoi v Tsentra'nom Lektorii Obshchestva v Moskve* (Pravda, 1949).

36. Among the many fine studies of Lysenko and Lysenkoism, see Joravsky, *The Lysenko Affair*; Medvedev, *The Rise and Fall of T. D. Lysenko*; Valerii Soifer, *Lysenko and the Tragedy of Soviet Science* (Rutgers University Press, 1994).

37. On the short-lived but fascinating effort to derail the establishment of cybernetics in the USSR, see Slava Gerovitch, *From Newspeak to Cyberspeak* (MIT Press, 2002).

38. Loren Graham, *Science, Philosophy and Human Behavior in the USSR* (Columbia University Press, 1987).

39. For discussion of the interaction of Nazi ideology and physics, see Mark Walker, *Nazi Science: Myth, Truth and the German Atomic Bomb* (Plenum, 1995).

40. In *Stalin and the Soviet Science Wars* (Princeton University Press, 2006), Ethan Pollock discusses the impact of Stalinism on physics, biology, economics, and a few other fields. See also *Tragicheskie Sud'by: Repressirovannye Uchenye Akademii Nauk SSSR*, ed. Viktor Kumanev (Nauka, 1995).

41. On Landau, see Karl Hall, "The Schooling of Lev Landau: The European Context of Postrevolutionary Soviet Theoretical Physics," *Osiris* 23 (2008): 230–259. See also Yuri Ranyuk et al., "Eshche raz ob 'Antisovetskoi zabostovke Khar'kovskikh fizikov,'" *Voprosy Istorii Estestvoznaniia i Tekhniki*, 2007, no. 3: 69–81. On the impact of Stalinism on the physics community in the 1930s, see Gennady Gorelik, "Moskva, Fizika, 1937 god," *Voprosy Istorii Estestvoznaniia i Tekhniki*, 1992, no. 1: 15–32.

42. S. P. Kapitsa, *Pis'ma o Nauke, 1930–1980* (Moskovskii Rabochii 1989); Lawrence Badash, *Kapitza, Rutherford and the Kremlin* (Yale University Press, 1985).

43. Gennedy Gorelik and Viktor Frenkel', *Matvei Petrovich Bronstein and Soviet Theoretical Physics in the Thirties* (Birkhäuser, 1994).

44. Robert A. McCutcheon, "The 1936–1937 Purge of Soviet Astronomers," *Slavic Review* 50 (1991), spring: 100–117.

45. V. L. Ginzburg, *O Nauke, o Sebe, i o Drugikh*, third edition (Fizmatlit, 2004), pp. 440–442.

46. Ibid., pp. 440–442.

47. Ibid., p. 443.

48. Ibid., p. 444.

49. Ibid., pp. 444–447.

50. A. S. Sonin, *Fizicheskii Idealizm: Istoriia Odnoi Ideologicheskoi Kampanii* (Fizmatlit, 1994); A. M. Blokh, *Sovetskii Soiuz v Inter'ere Nobelevskikh Premii* (Fizmatlit, 2005), pp. 321–331.

51. Viktor Frenkel, *Jacov Il'ich Frenkel: His Work, Life and Letters* (Birkhäuser, 1996); Iakov Ilich Frenkel', *Vospominaniia, Pis'ma, Dokumenty*, second edition (Nauka, 1986).

Chapter 3

1. http://nobelprize.org.nobelprizes/physics/laureates/2000/alferov-autobio.html.

2. International Conference on the Peaceful Uses of Atomic Energy, *Proceedings of the International Conference on the Peaceful Uses of Atomic Energy: Held in Geneva 8 August–20 August 1955* (United Nations, 1956), 16 volumes.

3. Walter McDougall, . . . *The Heavens and the Earth: A Political History of the Space Age* (Basic Books, 1985).

4. For an encyclopedic account of the movement, see Joshua Rubenstein, *Soviet Dissidents: The Struggle for Human Rights* (Beacon, 1985).

5. Paul Forman, "Behind Quantum Electronics: National Security as Basis for Physical Research in the United States, 1940–1960," *Historical Studies in the Physical and Biological Sciences* 18 (1987): 152–268.

6. On ONR, see Harvey Sapolsky, *Science and the Navy: A History of the Office of Naval Research* (Princeton University Press, 1990).

7. For an encyclopedic internal history of the US Atomic Energy Commission, see Richard G. Hewlett and Oscar E. Anderson Jr., *A History of the United States Atomic Energy Commission* (Pennsylvania State University Press, 1962 and 1969) (originally published as *The New World, 1939/1946* and *Atomic Shield, 1947/1952*). See also

Richard G. Hewlett and Jack Holl, *Atoms for Peace and War, 1953–1961: Eisenhower and the Atomic Energy Commission* (University of California Press, 1989).

8. In *A Peril and a Hope—Revised Edition: The Scientists' Movement in America, 1945–47* (MIT Press, 1971), Alice Kimball Smith describes these events and discusses the entry of scientists into the political arena over their concern that science "should henceforth be primarily an instrument of peace and dedicated to human welfare" (p. vi).

9. On McCarthyism and the scientists Bernard Peters and Philip Morrison, see Sam Schweber, *In the Shadow of the Bomb: Bethe, Oppenheimer and the Moral Responsibility of the Scientist* (Princeton University Press, 2000). See also B. J. Bernstein, "In the Matter of J. Robert Oppenheimer," *Historical Studies in the Physical Sciences* 12 (1982), no. 2: 192–252.

10. In *Unarmed Forces: The Transnational Movement to End the Cold War* (Cornell University Press, 1999), Matthew Evangelista discusses how Soviet and Western scientists and physicists debated scientific and policy issues of arms control, and in some cases convinced Soviet leaders to understand the need to modify their positions.

11. Susan Costanzo, "The 1959 Liriki-Fiziki Debate: Going Public with the Private?" in *Private Spheres of Soviet Russia*, ed. Lewis Siegelbaum (Palgrave 2006). My thanks to Professor Costanzo for sharing her work with me.

12. Translated by Ekaterina Chukaeva, English Department, Pomor State University, Arkhangelsk.

13. C. P. Snow, *The Two Cultures and the Scientific Revolution* (Cambridge University Press, 1959).

14. L. L. Kerber, *Stalin's Aviation Gulag: A Memoir of Andrei Tupolev and the Purge Era* (Smithsonian Institution Press, 1996).

15. http://nobelprize.org.nobelprizes/physics/laureates/2000/alferov-autobio.html

16. M. S. Sominskii, *Abram Fedorovich Ioffe* (Nauka, 1960), pp. 559–563.

17. Sominskii, *Ioffe*, pp. 566–571.

18. "Komar Anton Panteleymonovich: The First Head of the High-Energy Physics Laboratory at PNPI," http://hepd.pnpi.spb.ru/hepd/structure/people/komar.html

19. Viktor Frenkel, *Yakov Ilich Frenkel* (Birkhäuser, 1996), pp. 261–270, 285.

20. The following biographical section on Tuchkevich is drawn in part from *K 90-letiiu akademika Vladimira Maksimovicha Tuchkevicha*, ed. Zh. I. Alferov (FTI im. Ioffe, 1994).

21. Bertrand M. Patenaude, *The Big Show in Bololand: The American Relief Expedition to Soviet Russia in the Famine of 1921* (Stanford University Press, 2002).

22. On the contribution of Soviet scientists to the war effort, including that of LFTI physicists, see E. I. Grakina, *Uchenye—Frontu* (Nauka, 1989).

23. I. V. Grekhov, "K 100-letnemu iubileiu V. M. Tuckkevicha," *Istoricheskie khroniki,* pp. 32–33.

24. http://www.tiscali.co.ukreference/encyclopaedia/hutchinson/m0016610.html

25. Grekhov, "Tuckkevicha," pp. 32–36.

26. Ibid., p. 33.

27. Alferov, *Vladimira Maksimovicha Tuchkevicha,* p. 7.

28. A LFTI, F. 3, op. 1, ed. khr. 475, ll. 21–23.

29. A LFTI, F. 3, op. 1, ed. khr. 475, ll.29–33; Alferov et al., "Moshchnyi impul'snyi germanievyi vypriamitel'," *Izvestiia LETI im. V. I. Ul'ianova (Lenina)* 42 (xlii) (1960); Alferov et al., "Temperatura p-n perekhoda moshchnykh germanievykh ventilei," *Elektrichestvo,* 1962; "The Development of Germanium Diodes and Triodes That Replace Low-Power Radiolamps," "The Production of Large (no less than 5 cm^2) Rectifying Surfaces with the Goal of Establishing the Possibility of the Creation of Powerful Germanium Rectifiers at 200 amps and power of 100 volts," and "The Development of a semiconductor Gate Valve Instrument at 3,500 amps and 330 volts Using Germanium Rectifying Elements."

30. Alferov, *Vladimira Maksimovicha Tuchkevicha,* p. 8.

31. Grekhov, "Tuckkevicha," p. 33; Alferov, *Vladimira Maksimovicha Tuchkevicha,* pp. 11–12. For a compelling, even timeless discussion of the innovation process in Soviet industry, see Joseph Berliner, *Factory and Manager in the USSR* (Harvard University Press, 1957).

32. On Bardeen's family life, career path, and achievements in physics, see Lillian Hoddeson and Vicki Daitch, *True Genius: The Life and Science of John Bardeen* (Joseph Henry Press, 2002).

33. Alferov, *Vladimira Maksimovicha Tuchkevicha,* pp. 11–13.

34. Ibid., pp. 11–13.

35. Ibid., p. 9.

36. Grekhov, "Tuckkevicha," p. 35; Alferov, *Vladimira Maksimovicha Tuchkevicha,* pp. 9–10.

37. On the Efremov Scientific Research Institute of Electrophysical Apparatuses, see Josephson, *Red Atom,* pp. 192–194, 224–229, and http://www.niiefa.spb.su/. NIIEFA has a major role in proving equipment to the ongoing International Thermonuclear Experimental Reactor (ITER) project.

38. Grekhov, "Tuckkevicha," pp. 31, 35.

39. The bards' music circulated by means of magnitizdat (tapes of live recordings). The repeated copying and the low quality of the tape recorders meant that the quality of many copies was very low. This did not matter to the people who made the tapes at performances and circulated them among friends; musicians of all skill levels learned the songs, and performed them in homes of sympathizers, in cafés, and in parks. It was dangerous to perform the music or carry the tapes, since the authorities might arrest you, but that was part of the strange joy that the bards and their followers embraced. (Samizdat circulated by means of repeated typing, using several sheets of onionskin and carbon paper, so its quality was also very low.).

40. Interview with Zhores Alferov, St. Petersburg, NOTs, October 24, 2007. Apparently only Okudzhava performed at the LFTI in 1968; Galich never performed there.

41. Arseny Berezin, *Piki-Kozyri* (Pushkin Fund, 2007), pp. 204–211.

42. Josephson, *New Atlantis Revisited*, pp. 296–302.

43. While Akademgorodok maintained its special status as an oasis of science and culture in the USSR, it became grayer, like the rest of the USSR. Recently, in memory of its special place in Soviet history and of the special role of the bards, scientists connected with the social club "Under the Integral" hung a plaque dedicated to Galich on the building in which they met in the 1960s. Then a dining hall, it is now a bank.

44. "The Trial Begins," *Time*, February 18, 1966.

45. Grekhov, "Tuckkevicha," p. 35.

46. Ibid., pp. 31, 35.

47. See, for example, G. V. Romanov, "Vysokii dolg kommunista," and "Leninskim kursom—K novym pobedam v kommunisticheskom stroitel'stve," in his *Izbrannye Rechi i Stat'i* (Izdatpolit, 1980), pp. 243–258 and 479–501.

48. Grekhov, "Tuckkevicha," pp. 33–35. On the complex relationship between politics, science, and industry in the USSR, see *Industrial Innovation in the Soviet Union*, ed. Ronald Amann and Julian Cooper (Yale University Press, 1982). On the "Intensification-90" program, see L. N. Zaikov, "Povyshat' uroven' partiinogo rukovodstva komsomolom," *Leningradskaia pravda*, July 21, 1984, pp. 1–2; "V Tsentral'nom komitete KPSS o rabote provodimoi Leningradskim obkomom KPSS po usileniiu intensifikatsii ekonomiki v dvenadtsatoi piatiletke na osnove uskoreniia nauchnotekhnicheskogo progressa," ibid., August 4, 1984, p. 1; "V obkome KPSS, 'Intensifikatsiia-90' programma deistviia po povysheniiu effektivnosti ekonomiki," ibid., August 7, 1984, p. 1.

49. Grekhov, "Tuckkevicha," pp. 31–35.

50. For Alferov's reminiscences of Alexandrov, see Alferov, *Nauka i Obshchestvo*, pp. 190–194.

51. For a sense of Aleksandrov's attitudes toward science and politics in a series of speeches and published articles, see his *Nauka—Strane: Stat'i i vystupleniia* (Nauka, 1983).

52. Boris Zakharchenia, "Nebol'shaia Saga o Zhorese Alferove," in Zh. I. Alferov, *Fizika i Zhizn'*, second edition (Nauka, 2001), pp. 15–16.

53. Ginzburg, "Protiv biurokratizma, perestrakhovki i nekompetenentnost'," in *I Nogo ne Dano*, ed. Iu. N. Afanas'ev (Progress, 1988).

54. Email from Arseny Berezin, former LFTI employee, May 13, 2009.

55. Ernest Braun, "Selected Topics from the History of Semiconductor Physics and Its Applications," in *Out of the Crystal Maze: Chapters from the History of Solid-State Physics*, ed. Lillian Hoddeson et al. (Oxford University Press, 1992), pp. 475–476. One-hundred twenty-one military scientists, 41 university scientists, and 139 industrial scientists attended the first symposium, the published proceedings of which amounted to nearly 800 pages.

56. Interview with Zhores Alferov, January 17, 2009, St. Petersburg.

57. Ibid.

58. Harvey Sapolsky, *Science for the Navy: The History of the Office of Naval Research* (Princeton University Press, 1990).

59. Arkadii Sosnov, "Moe Otkrytie Ameriki," *Sankt-Peterburgskie Vedomosti*, June 11, 2003, p. 5.

60. Hoddeson and Daitch, *True Genius*, pp. 275–277.

61. Zakharchenia, "Alferov," pp. 15–16.

62. Alferov, Interview, St. Petersburg, January 17, 2009.

63. Yakov Rabkin, *Science Between the Superpowers* (Priority, 1988).

64. Interview with Zhores Alferov, St. Petersburg, NOTs, October 24, 2007.

65. Hoddeson and Daitch, *True Genius*, pp.173–177.

66. See Nick Holonyak, "John Bardeen and the Point-Contact Transistor," *Physics Today* 45 (1992), no. 4: 36–43; Frederick Nebeker, "Interview with Nick Holonyak," June 22, 1993, available at www.pbs.org.

67. N. Holonyak Jr. et al., ""Luminescence, Lasers, Carrier Interaction Effects, and Instabilities in Compound and Elemental Semiconductors," in *Summary of Engineering Research* (University of Illinois, 1971), RS 11/2/805, University of Illinois Archives, p. 112.

68. A LFTI, F. 3, op. 1, ed. Khr. 618, "Otchet o komandirovke v SShA zav. Sektorom FTI im. Ioffe AN SSSR doktora fiz.-mat. Nauk Zh. I. Alferov," ll. 25–28.

69. Kressler focused on semiconductor detectors in optical communications, while Rappaport focused on the materials and science of photoelectricity. See Paul Rappaport, "Photoelectricity," *Proceedings of the National Academy of Sciences of the United States of America* 47 (1961), no. 8: 1303–1306.

70. A LFTI, F. 3, op. 1, ed. Khr. 618, ll. 25–37.

71. Ibid., ll. 25–28.

72. Arkadii Sosnov, "Moe Otkrytie Ameriki," *Sankt-Peterburgskie Vedomosti,* June 11, 2003, p. 5.

73. A LFTI, F. 3, op. 1, ed. khr. 710.

74. A LFTI, F. 3, op. 1, ed. khr. 475. Presentation for Consideration for Lenin Prize (1965) for "Research of Complex Structures With p-n heterojunctions, the Development of Technology of Manufacture and Assimilation into Serial Production of Powerful Silicon Gates/Rectifiers that Guarantee the Development and Production of Devices on their Basis." See also V. M. Tuchkevich "Silovye poluprovidnikovye ventili," *Vestnik Akademii Nauk,* 1964, no. 9; Tuchkevich, *Neupravlaiemye silovye ventili* (Znanie RSFSR, 1964).

75. A LFTI, F. 3, op. 1, ed. khr. 475, ll. 2–6. The nominations for prizes had to include assertions that they contained no trade or state secrets or restricted information, but could be published on TV or radio. See ibid., ll. 34, 36–37.

76. A LFTI, F. 3, op. 1, ed. khr. 475, ll. 24–25.

77. Ibid., ll. 75–77.

78. Ibid., ll. 73–74.

79. Ibid., ll. 29–33; Alferov et al., "Moshchnyi impul'snyi germanievyi vypriamitel'," *Izvestiia LETI im. V. I. Ul'ianova (Lenina)* 42 (xlii) (1960); Alferov et al., "Temperatura p-n perekhoda moshchnykh germanievykh ventilei," *Elektrichestvo,* 1962.

80. "O prisuzhdenii Leninskikh premii 1972 goda v oblasti nauki i tekhniki," *Pravda,* April 22, 1972, pp. 1, 3.

81. A LFTI, F. 3, op. 1, ed. Khr, 594, ll. 295–316 (Alferov, "Otchet o poezdke v Vengriiu na Mezhdunarodniu konferentsiiu po fizike i khimii poluprovodnikovykh geteroperekhodov i sloistykh struktur, 12–17 October 1970"); Iu. Koptev, "Pobeda fizikov," *Leningradskaia Pravda,* April 30, 1972.

82. "Novoe popolnenie Akademii," *Leningradskaia Pravda,* March 24, 1979.

83. V. Kulik-Remezova, "Molodost' nauki," *Smena,* November 3, 1979, pp. 1, 4. Local reporters frequently wrote about the LFTI, its physicists, and their great achievements. See for example, "Krupneishii tsentr fizicheskoi nauki," *Leningradskaia Pravda,* June 10, 1979.

84. Kulik-Remezova, "Molodost' nauki," pp. 1, 4.

85. Interview with Semen Grigorievich Konnikov, LFTI, St. Petersburg, January 16, 2007.

Chapter 4

1. My deep thanks to Dr. Mort Panish for his extensive critical comments on this section. Dr. Spencer Weart also made important suggestions.

2. M. S. Sominskii, *Abram Fedorovich Ioffe* (Nauka, 1960), pp. 559–563.

3. http://en.wikipedia.org/wiki/Heterojunction

4. Boris Zakharchenia, "Nebol'shaia saga o Zhorese Alferove," in *Fizika i Zhizn'*, ed. Zh. I. Alferov (Nauka, 2001), pp. 9–10.

5. Zhores Alferov, "The Double Heterostructure: Concept and Its Applications in Physics, Electronics and Technology," Nobel Lecture, December 8, 2000 (hereafter cited as Alferov, Nobel Lecture), p. 414. See also N. G. Basov, O. N. Krokhin, and Iu. M. Popov, "The Possibility of Use of Indirect Transitions to Obtain Negative Temperature in Semiconductors," *Soviet Physics—JETP* 12 (1961), May: 1033; D. N. Nasledov et al., "Recombination Radiation of Gallium Arsenic," *Fizika Tverdogo Tela* 4 (1962): 1062–1065; R. H. Hall et al., "Coherent Light Emission from GaAs Junction," *Physical Review Letters* 9 (1962): 366–378; N. Holonyak Jr. and S. F. Bevacqua, "Coherent (Visible) Light Emission from Ga(As$_{1-x}$P$_x$) Junctions," *Applied Physics Letters* 1 (1962), no. 4: 83–84.

6. http://en.wikipedia.org/wiki/Semiconductor

7. On the history of the people who advanced the theoretical and experimental program of solids state physics, see *Out of the Crystal Maze: Chapters from the History of Solid-State Physics*, ed. Lillian Hoddeson et al. (Oxford University Press, 1992). (Alferov, Holonyak, Kroemer, and Kilby are not mentioned, nor are integrated circuits or heterojunctions.).

8. Paul Forman, "Behind Quantum Electronics: National Security as Basis for Physical Research in the United States, 1940–1960," *Historical Studies in the Physical and Biological Sciences* 18 (1987): 152–268.

9. Lillian Hoddeson and Vicki Daitch, *True Genius: The Life and Science of John Bardeen* (Joseph Henry Press, 2002), pp.172–173.

10. Hoddeson, *Out of the Crystal Maze*, pp. 202–208.

11. Ernest Braun, "Selected Topics from the History of Semiconductor Physics and Its Applications," in Hoddeson, *Out of the Crystal Maze*, pp. 443–445.

12. Ibid., pp. 446–447.

13. Ibid., pp. 458–459.

14. Andrew Tickle, *Thin-Film Transistors: A New Approach to Microelectronics* (Wiley, 1969), pp. 1–2.

15. V. Rykunov, "Perekhod, vedushchii v zavtra," *Sotialisticheskaia Industriia*, May 7, 1972.

16. In this section I draw heavily on Richard Fitzgerald, "Physics Nobel Prize Honors Roots of Information Age," *Physics Today* 53 (2000), no. 12: 17–19.

17. J. S. Kilby, *IEEE Transactions on Electron Devices*, Vol. E-D 23 (1976) p. 648.

18. Tickle, *Thin-Film Transistors*, pp. 1–2.

19. H. Kroemer, *Proceedings of Institute of Radio Engineers* 45 (1957), p. 1535.

20. Manfred Thumm, "Historical German Contributions to Physics and Applications of Electromagnetic Oscillations and Waves," http://www.radarworld.org/history.pdf.

21. See US patent 3,309,553 (filed Aug. 16, 1963, issued 1967). Also see Varian Central Research Report CRR-36, 1963 (unpublished, but available from Herbert Kroemer in pdf).

22. Thumm, "Historical German Contributions.

23. Interview with Vsevolod Lundin, LFTI, St. Petersburg, January 16, 2007.

24. The latter materials are not naturally lattice matched, and the correct ratios of Ga and In and of As and P must be used to get lattice matching.

25. R. N. Hall, *IEEE Transactions on Electron Devices*, Vol. ED-23 (1976), pp. 674–678. On G. N. Hall, see http://www.semiconductormuseum.com/Transistors/GE/Oral Histories/Hall/HallIndex.htm.

26. H. C. Casey Jr. and M. B. Panish, *Heterostructure Lasers. Part A. Fundamental Principles* (Academic, 1978), p. 3.

27. D. N. Nasledov, A. A. Rogachev, S. M. Ryvkin, V. E. Khartsiev, and B. V. Tsarenkov, *Fizika Tverdogo Tela* 4 (1962), pp. 1062 and 3346.

28. H. Kroemer, "A Proposed Class of Heterojunction Injection Lasers," *Proceedings of IEEE* 51 (1963), pp. 1782–1783; Zh. I. Alferov, and R. F. Kazarinov, Authors' Certificate 28448 (USSR), as cited in Zh. I. Alferov, D. Z. Garbuzov, V. S. Grigor'eva. Yu. V. Zhilyaev, L. V. Kradinova, V. I. Korol'kov, E. P. Morozov, O. A. Ninua, E. L. Portnoi, V. D. Prochukhan, and M. K. Trukan, *Fizika Tverdogo Tela* 9 (1967), 279; *Soviet Physics—Solid State*, 1967: 208. See also Zh. I. Alferov, *Physica Scripta* 68 (1996), p. 32; Kroemer, *Physica Scripta* 68 (1996), p. 10; Yurii M. Popov, "On the history of the invention of the injection laser," *Physics-Uspekhi* 47 (2004): 1068–1070; Casey and Panish, *Heterostructure Lasers. Part A*, pp. 5–6; *Physics of p-n Junctions and Semiconductor Devices*, ed. S. M. Ryvkin and Iu. V. Shmartsev (Consultants Bureau, 1971), p. 287.

29. Alferov, Nobel Lecture, p. 414. See also Alferov and R. F. Kazarinov, "Semiconductor Laser with Electric Pumping," Inventor's Certificate No. 181737 (in Russian), Application 950840, March 30, 1963; H. Kroemer, "A Proposed Class of Heterojunction Injection Lasers," *Proceedings of IEEE* 51 (1963), p. 1782.

30. Email, Mort Panish to Paul Josephson, May 18, 2009.

31. Herbert Kroemer, "Heterostructures for Everything: Device Principle of the 1980s?" *Japanese Journal of Applied Physics* 20 (1981), Supplement 20–1, pp. 9–13; Zhores Alferov, *Physica Scripta* 68 (1996), p. 32.

32. http://www.ieee.org/web/aboutus/historycenter/biography/kroemer.html

33. Alferov, Nobel Lecture, p. 415. See also Zh. I Alferov, V. B. Khalfin, and R. V. Kazarinov, "A Characteristic Feature of Injection into Heterojunctions," *Fizika Tverdogo Tela* 8 (1966): 3102–3105. Also in 1966, Alferov summarized the understandings of the main advantages of DHS for lasers, and high power rectifiers in particular, in "Possible Development of a Rectifier for Very High Current Densities on the Bases of p-I-n (p-n-n+, n-p-p+) Structure with Heterojunctions," *Fizika i Tekhnika Poluprovodnikov* 1: 436–438.

34. Alferov, Nobel Lecture, p. 416. See also R. L. Anderson, "Germanium-Gallium Arsenide Heterojunctions," *IBM Journal of Research and Development* 4 (1960), p. 263; Anderson, "Experiments on Ge-GaAs Heterojunctions," *Solid State Electronics* 5 (1962), p. 341.

35. Interview with Alferov, January 17, 2009, St. Petersburg.

36. Rykunov, "Perekhod, vedushchii v zavtra."

37. Zakharchenia, "Nebol'shaia saga o Zhorese Alferove," pp. 13–14.

38. Zh. I. Alferov et al., *Soviet Physics—Semiconductors* 4 (1971): 1573; I. Hayashi, M. Panish, P. W. Foy, and S. Sumski, *Applied Physics Letters* 17 (1970): 109; Mahmoud Manasreh, *Semiconductor Heterojunctions and Nanostructures* (McGraw-Hill, 2005), p. 452.

39. Rykunov, "Perekhod, vedushchii v zavtra."

40. Alferov, Nobel Lecture, pp. 416–417. See also Alferov et al., "High Voltage p-n Junctions in $Ga_xAl_{1-x}As_{1-x}$," *Fizika i Tekhnika Poluprovodnikov* 5 (1971), p. 196; H. S. Rupprecht, J. M. Woodall, and G. D. Pettit, "Efficient Visible Electroluminence at 300 K from $Ga_xAl_{1-x}As_{1-x}$ p-n Junctions Grown by Liquid-Phase Epitaxy," *Applied Physics Letters* 11 (1967), p. 81; Casey and Panish, *Heterostructure Lasers. Part A*, pp. 6–7; H. Nelson, *RCA Review* 24 (1963), 603.

41. Zh. I Alferov et al., "Heterojunctions on the Base of A B Semiconducting and of their Solid Solutions," in *International Conference on the Physics and Chemistry of*

Semiconductor Heterojunctions and Layer Structures, ed. I. G. Szigeti (Academiai Kiado, 1971); Alferov, Nobel Lecture, p. 419.

42. Email, Mort Panish to Paul Josephson, May 18, 2009.

43. Casey and Panish, *Heterostructure Lasers. Part A*, pp. 7–8. See also H. Kressel and H. Nelson, "Close-Confinement Gallium Arsenide P-N Junction Lasers with Reduced Optical Loss at Room Temperature," *RCA Review* 30 (1969): pp.106–13, 1969; I. Hayashi and M. B. Panish, *Journal of Applied Physics* 41 (1970), p. 150. In an email message to Paul Josephson dated May 18, 2009, Panish wrote: "I think we were aware of Kressel and Nelson's work and if my memory is correct . . . we thought they were the major competition and we weren't particularly worried by them."

44. I. Hayashi, "Heterostructure Lasers," *IEEE Transactions on Electron Devices* ED-31 (1984), p. 1645. See also Alferov, Nobel Lecture, p. 418.

45. Zh. I Alferov et al., "Investigation of the Influence of AlAs-GaAs Heterostructure Parameters on the Laser Threshold Current and the Realization of Continuous Emission at Room Temperature," *Fizika i Tekhnika Poluprovodnikov* 4 (1970): 1826–1829; Alferov, Nobel Lecture, p. 418.

46. I. Hayashi et al., "Junction Lasers Which Operate Continuously at Room Temperature," *Applied Physics Letters* 17 (1970), pp. 109–111; Alferov, Nobel Lecture, p. 418.

47. Interview with Dr. Mort Panish, January 31, 2009, Freeport, Maine.

48. Casey and Panish, *Heterostructure Lasers. Part A*, pp. 8–9. See also Zh. I. Alferov, V. M. Andreev, V. I. Korol'kov, E. I. Portnoi, and D. N. Tret'yakov, *Fizika i Tekhnika Poluprovodnikov* 2 (1968), p. 1016; *Soviet Physics—Semiconductors* 2 (1969), 843; Zh. I. Alferov, V. M. Andreev, E. L. Portnoi, and M. K. Trukan, *Fizika i Tekhnika Poluprovodnikov* 3 (1969), 1328; *Soviet Physics—Semiconductors* 3 (1970), p. 1107; Zh. I. Alferov, V. M. Andreev, D. Z. Garbuzov, Yu. V. Zhilyaev, E. P. Morozov, E. L. Portnoi, and V. G. Trofim, *Fizika i Tekhnika Poluprovodnikov* 4 (1970), 1826; *Soviet Physics—Semiconductors* 4 (1971), p. 1573. In an email message to me dated May 18, 2009, Panish wrote: "I don't know what you do with this, but we almost certainly had CW lasing at least six weeks earlier. However, Hayashi insisted that we must have the lasing spectrum and our early lasers didn't last long enough to plot the spectrum with the slow equipment we had at the time. I don't think Alferov bothered to show a spectrum in his first CW paper. He took the current-light output plot as evidence enough as I remember. He was right. Hayashi didn't think that was sufficient so we waited until we had a laser that lived long enough."

49. M. A. Novikov, "Oleg Vladimirovich Losev: Pioneer of Semiconductor Electronics," *Physics of the Solid State* 46 (2004), no. 1: 1–4.

50. *Handbook of Thin Film Technology*, ed. Leon Maissel and Reinhard Gland (McGraw-Hill, 1970), pp. 1–1, 2–2, 10–2.

51. Manasreh, *Semiconductor Heterojunctions*, pp. 329–330.

52. http://www.siliconfareast.com/epitaxy.htm

53. Iu. Koptev, "Lenty iz Poluprovodnikov," *Izvestiia*, March 24, 1972.

54. *Handbook of Thin Film Technology*, pp. 1–1, 2–2, 10–2.

55. Tickle, *Thin-Film Transistors*, pp. 2–3. See also P. K. Weimer et al., "A 180-Stage Integrated Thin-Film Scan Generator," *Proceedings of IEEE* 5 (1966), March: 359–360.

56. Tickle, *Thin-Film Transistors*, pp. 18–19.

57. Alferov, Nobel Lecture, p. 424.

58. Interview with Vsevolod Lundin in laboratory at LFTI, St. Petersburg, January 16, 2007.

59. Interview with Vsevolod Lundin in laboratory at LFTI, St. Petersburg, January 16, 2007. See also www.aixtron.com.

60. Interview with Panish, January 31, 2009.

61. On achievements in optical computing, optical processing components, optical signal processing, and optical image processing in the former Soviet Union in the early 1990s, see *Proceedings of Second International Conference on Optical Information Processing*, ed. Zh. I Alferov, Y. Gulyaev, and D. Pape. On these devices, see Manasreh, *Semiconductor Heterojunctions*, pp. 457–448.

62. Fitzgerald, "Physics Nobel Prize Honors Roots of Information Age," *Physics Today* 53, no. 12: 17.63. Zh. I Alferov, "Photovoltaic Solar Energy Conversion," pp. 11–15. Neither Alferov nor his secretary nor I can find the full reference for this source.

63. Ibid., pp. 11–15.

64. Ibid., pp. 19–25.

65. Robert Service, "Can the Upstarts Top Silicon?" *Science* 319 (2008), February 8: 718–720.

66. Alferov, "Photovoltaic Solar Energy Conversion," pp. 26–32. Of course, the manufacturing of photocells with such materials as gallium arsenide creates another set of pollution problems.

Chapter 5

1. David Remnick's *Lenin's Tomb* (Random House, 1993) remains the most engaging treatment of the Gorbachev era, glasnost, and perestroika.

2. For the politics and economics of muddling through in the Brezhnev era, see Valerie Bunce, "The Political Economy of the Brezhnev Era: The Rise and Fall of

Corporatism," *British Journal of Political Science* 13 (1983), no. 2: 129–158; Donald Kelly, ed., *Soviet Politics in the Brezhnev Era* (Praeger, 1980); Alfred Evans Jr., "Developed Socialism in Soviet Ideology," *Soviet Studies* 29 (1977), no. 3: 409–428; T. H. Rigby, "The Soviet Regional Leadership: The Brezhnev Generation," *Slavic Review* 37 (1978), no. 1: 1–24; Sidney Ploss, "Soviet Politics on the Eve of the 24th Party Congress," *World Politics* 23 (1970), no. 1: 61–82; Seweryn Bialer, *Stalin's Successors: Leadership, Stability, and Change in the Soviet Union* (Cambridge University Press, 1980). For a sense of Brezhnev's concerns from the man himself, see Robert Neal's review of Brezhnev's "Collected Works," *Slavic Review* 36 (1977), no. 3: 488–493.

3. For a discussion of this investment dilemma, see Boris Rumer, *Investment and Reindustrialization in the Soviet Economy* (Westview, 1984).

4. Howell Raines, "Thatcher's Visit: Glasnost in Action?" *New York Times,* April 3, 1987.

5. For one of many good studies on the man, see Archie Brown, *The Gorbachev Factor* (Oxford University Press, 1996).

6. See Jack Matlock, *Reagan and Gorbachev: How the Cold War Ended* (Random House, 2004).

7. Email from Stephen Fortescue to the author, July 3, 2009.

8. Ibid.

9. For a brief summary of debates surrounding Siberian river diversion, see Josephson, *New Atlantis Revisited* (Princeton University Press, 1997), pp. 185–197.

10. For example, M. G. Iaroshevskii, ed., *Repressirovnaia Nauka* (Nauka, 1991).

11. Iu. N. Afanas'ev, ed., *Inogo ne Dano* (Moscow: Progress, 1988).

12. V. Ginzburg, "Protiv biurokratizma, perestrakhovki i nekompetenentnost'," in *Inogo ne Dano,* ed. Afanas'ev, pp. 136–139.

13. Ibid., p. 139.

14. Ibid., pp. 140–141.

15. Ibid., pp. 139–144. The only other Soviet physicist to win the medal, the nuclear specialist Georgii Flerov, is known also for having written Stalin in 1942 about the need for the USSR to commence an atomic bomb project.

16. S. Samoilis, "Ne Otkladyvaia na zavtra," *Leningradskaia Pravda,* December 6, 1989.

17. Zh. I. Alferov, "Imet' provintsial'nuiu nauku—bessmyslenno," *Izvestiia,* November 10, 1989.

18. E. A. Tropp, "Peterburgskaia Nauka: Devianostye Gody XX v.," in *Akademiches-kaia Nauka v Sankt-Peterburge v XVIII-XX vekakh: Istoricheskie Ocherki*, ed. Zh. I. Alferov (Nauka, 2003), pp. 538–539.

19. Samoilis, "Ne Otkladyvaia na zavtra," p. 2.

20. Archive of LFTI (hereafter A LFTI), Tuchkevich to Marchuk, October 21, 1986.

21. Blair Ruble, *Shaping a Soviet City* (University of California Press, 1990), p. 116. See also (cited in Ruble) L. N. Zaikov, "Povyshat' uroven' partiinogo rukovodstva komsomolom," *Leningradskaia Pravda*, July 21, 1984, pp. 1–2; "V Tsentral'nom komitete KPSS o rabote provodimoi Leningradskim obkomom KPSS po usileniiu intensifikatsii ekonomiki v dvenadtsatoi piatiletke na osnove uskoreniia nauchnotekhnicheskogo progressa," ibid., August 4, 1984, p. 1; "V obkome KPSS, 'Intensifikatsiia-90' programma deistviia po povysheniiu effektivnosti ekonomiki," ibid., August 7, 1984, p. 1; "Nastoichivo dvizat'sia vpered: Vystuplenie M. S. Gorbacheva na sobranii aktiva Leningradskoi partiinoi organizatsii 17 maia 1985 goda," *Kommunist*, 1985, no. 8:23–34; and "Soedinenie nauki s proizvodstvom velenie vremeni: Prebyvanie M. S. Gorbacheva v Leningrade," *Leningradskaia Pravda*, May 17, 1985, p. 1.

22. See also A LFTI, V. M. Tuchkevich to E. P. Velikhov, March 11, 1986.

23. Zh. I. Alferov, "Bez nauki net perspektivy," *Zvezda*, 1989, no. 3, pp. 162–163.

24. Ibid., pp. 162–163.

25. Ibid., p. 165.

26. Ibid., pp. 162–163.

27. Ibid., pp. 162–163.

28. A LFTI, Alferov to Prokhorov, December 6, 1988.

29. A LFTI, Alferov to V. Ia. Titov, July 24, 1988.

30. Vasily Grossman, *A Writer at War: Vasily Grossman with the Red Army, 1941–1945*, ed. Antony Beevor and Luba Vinogradova (Pantheon Books, 2006).

31. Zh. I. Alferov and E. A. Tropp, "Sankt-Peterburgskii Nauchnyi Tsentr—Istoricheskoe iadro Rossiiskoi Akademii Nauk," in E. A. Tropp, *Sankt-Peterburgskii Nauchnyi Tsentr Rossiiskoi Akademiii Nauk*, second edition (Nauka, 2003), pp. 3–7.

32. Alferov, "Bez nauki net perspektivy," pp. 165–166; Alferov and Tropp, "Sankt-Peterburgskii Nauchnyi Tsentr," pp. 3–7.

33. See e.g., I. A. Glebov and I. I. Sigov, eds., *Sovershenstvovanie Upravleniia Fundamental'nymi Issledovaniiami v Krupnom Gorode* (Nauka—Leningradskoe otdelenie, 1983).

34. Alferov, "Bez nauki net perspektivy," pp. 165–166.

35. Ibid., pp.165–166; Alferov and Tropp, "Sankt-Peterburgskii Nauchnyi Tsentr," pp. 3–7.

36. As John Heilbron has shown, Max Planck sought to avoid political pronouncements under Kaiser Wilhelm during World War I, the Weimar Republic of the 1920s, and the rise of the National Socialists even as he clearly occupied a central position as spokesman for the scientific establishment with three different governments. How could his silence on such issues as the role of German scientists during the war and the firing of Jewish colleagues be apolitical? See John Heilbron, *The Dilemmas of an Upright Man* (University of California Press, 1986).

37. Fang Lizhi, *Bringing Down the Great Wall: Writings on Science, Culture and Democracy in China* (Knopf, 1981).

38. See Andrei Sakharov, *Memoirs* (Knopf, 1990), on which this section is largely based. Jay Bergman's intellectual biography *Meeting the Demands of Reason: The Life and Thought of Andrei Sakharov* (Cornell University Press, 2009) adds considerably to our understanding of the evolution of Sakharov's scientific and political thinking. See also Gennadii Gorelik, *The World of Andrei Sakharov: A Russian Physicist's Path to Freedom* (Oxford University Press, 2005); Richard Lourie, *Sakharov: A Biography* (University Press of New England, 2002).

39. Matthew Evangelista, *Unarmed Forces* (Cornell University Press, 1999), pp. 56–59.

40. Academician Zeldovich refused to sign the letter, claiming that it should have been written in a different spirit and that he was thinking of preparing his own private response. Iulii Khariton, Sakharov's boss at Arzamas-16 (and his friend), believed that such a letter had to be dispatched because members of the Academy of Sciences, including himself, had already protested against actions of Academician Sakharov. Petr Kapitsa believed that it was necessary to invite Sakharov for a discussion at a meeting of the Presidium of the Academy of Sciences of the USSR and only then to decide upon the appropriate reaction to his conduct, but pointedly reminded his fellow academicians that the Berlin Academy had ousted Albert Einstein after Hitler took power. The economist Vladimir Kantorovich considered it inopportune to sign a collective letter since he was as a recent Nobel laureate and was thinking about making an individual protest. Vitalii Ginzburg, Sakharov's colleague at FIAN, refused to go along.

41. http://www.aip.org/history/sakharov/humrt.htm

42. Interview with Zhores Alferov, St. Petersburg, March 7, 2008.

43. Bill Keller, "Sakharov Nominated for Soviet Congress," *New York Times*, January 21, 1989, and "Soviet Academy Elects Sakharov to Legislature," *New York Times*, April 21, 1989.

44. Robert Horvath, *The Legacy of Soviet Dissent: Dissidents, Democratization and Radical Nationalism in Russia* (Routledge Curzon, 2005), pp. 106–116.

45. A. D. Sakharov, "Neizbezhnost' perestroiki," in *Inogo ne dano*, ed. Afanas'ev, pp. 122–135. Other contributors to this volume were energetic supporters of perestroika.

46. Vera Tolz, *The USSR's Emerging Multiparty System* (Praeger, 1990).

47. Interview with Zhores Alferov, St. Petersburg, March 7, 2008.

48. Interview with Zhores Alferov, St. Petersburg, NOTs, October 24, 2007.

49. Interview with Zhores Alferov, St. Petersburg, NOTs, October 24, 2007.

50. Tropp, "Peterburgskaia Nauka: Devianostye Gody XX v.," in *Akademicheskaia Nauka*, ed. Zh. I. Alferov, pp. 535–536.

51. Ibid., pp. 535–536.

52. Fifteen years later he voiced the opinion that, because of the breakup of the USSR, "tens of millions have suffered." He exaggeratedly claimed: "Twenty-five million Russians have gone abroad. After the Russian Revolution only 2.5 million people departed." Interview with Alferov, St. Petersburg, NOTs, October 24, 2007.

53. Alferov, "Vystuplenie deputata Alferova Zh. I.," *Leningradskaia Pravda*, December 19, 1989, pp. 1, 3. Also published in Alferov, *Fizika i Zhizn'* (Nauka, 2001), pp. 221–224.

54. Alferov, "Vystuplenie," pp. 1, 3.

55. Ibid., pp. 1, 3.

56. Interview with Zhores Alferov, St. Petersburg, January 13, 2006.

57. Frank N. von Hippel, "Better active today than radioactive tomorrow," http://inesap.org/bulletin22/bul22art18.htm

58. Vladislav Zubok, "Gorbachev's Nuclear Learning: How the Soviet Leader Became a Nuclear Abolitionist," http://bostonreview.net./BR25.2/zubok.html

59. Roald Sagdeev, *The Making of a Soviet Scientist: My Adventures in Nuclear Fusion and Space from Stalin to Star Wars* (Wiley, 1994).

60. Alferov, "Bez nauki net perspektivy," pp.167–168.

61. Ibid., pp. 167–168.

Chapter 6

1. Loren Graham and Irina Dezhina, *Science in the New Russia* (Indiana University Press, 2008), p. 37.

2. Ibid., p. 8.

3. Ibid., p. 3.

4. Alferov, quoted in *Neva Times*, September 18, 1993.

5. For a comprehensive discussion of Yeltsin's life and politics, see Timothy Colton, *Yeltsin: A Life* (Basic Books, 2008). On the Yeltsin era, see also *The New Russia: Troubled Transformation*, ed. Gail Lapidus (Westview, 1995); Michael McFaul, "State Power, Institutional Change, and the Politics of Privatization in Russia," *World Politics* 47, no. 2 (January 1995): pp. 210–243.

6. On US-Russian efforts to inventory and safeguard nuclear materials, see Graham Allison et al., *Avoiding Nuclear Anarchy* (MIT Press, 1996). Also see Richard Lugar, "The Republican Course," *Foreign Policy* 86 (1992), spring: 86–98.

7. On the challenges in introducing market reforms in Russia, including the ill-fated 500-day plan, see Padma Desai, "Russian Retrospectives on Reforms from Yeltsin to Putin," *Journal of Economic Perspectives* 19 (2005), no. 1: 87–106.

8. On Chechnya, the failure of Russian policy, and the great human costs of this war, see Anna Politkovskaia, *A Dirty War: A Russian Reporter in Chechnya* (Harvill, 2001). See also Carlotta Gall and Thomas de Waal, *Chechnya: Calamity in the Caucasus* (New York University Press, 1998); Anne Nivat, *Chienne de Guerre: A Woman Reporter Behind the Lines of the War in Chechnya* (Public Affairs, 2001); Michael McFaul, "Eurasia Letter: Russian Politics after Chechnya," *Foreign Policy* 99 (1995), summer: 149–165.

9. Graham and Dezhina, *Science in the New Russia*, pp. 10–16; E. A. Tropp, "Peterburgskaia Nauka: Devianostye Gody XX v.," in *Akademicheskaia Nauka v Sankt-Peterburge v XVIII-XX vekakh: Istoricheskie Ocherki*, ed. Zh. I. Alferov (Nauka, 2003), pp. 544–546.

10. Graham and Dezhina, *Science in the New Russia*, pp. 20–21; Tropp, "Peterburgskaia Nauka," p. 587.

11. Tropp, "Peterburgskaia Nauka," pp. 550–551.

12. Ibid., pp. 548–549.

13. Ibid., p. 540.

14. Paul Josephson and Irina Dezhina, "The Slow Pace of Reform of Science in Russia and Ukraine," *Problems of Post-Communism* 45 (1998), no. 5: 48–63; Paul Josephson and Igor Egorov, "Ukraine's Declining Scientific Establishment," *Problems of Post Communism* 49 (2002), no. 4: 43–51.

15. *Vechernii Peterburg*, February 18, 1993.

16. On INTAS's activities, see http://www.intas.be/. INTAS was liquidated on January 1, 2007.

17. On the International Science and Technology Centers from the perspective of the first director, and on the effort to support scientists of the former Soviet Union with weapons expertise in their research at home, see Glenn Schweitzer, *Moscow-DMZ* (M. E. Sharpe, 1996).

18. On the history of the International Science Foundation, see Irina Dezhina, *The International Science Foundation* (ISF, 2000).

19. St. Petersburg newspaper (misplaced by PRJ), November 21, 1993.

20. Arkadii Sosnov, "Moe Otkrytie Ameriki," *Sankt-Peterburgskie Vedomosti*, June 11, 2003, p. 5.

21. Interview with Zhores Alferov, NOTs, St. Petersburg, October 24, 2007.

22. Graham and Dezhina speak about the peer review reform throughout *Science in the New Russia*.

23. Graham and Dezhina, *Science in the New Russia*, pp. 46–54.

24. *Nauka v Rossii*, ed. S. F. Olden'burg (Gosizdat, 1923).

25. On the Russian Association of Physicists, see Josephson, *Physics and Politics in Revolutionary Russia* (University of California Press, 1991).

26. Tropp, "Peterburgskaia Nauka," pp. 551–552.

27. *Johnson's Russia List*, no. 5280 (June 2, 2001), at http://www.cdi.org/russia/john son/5280.html##5

28. Ibid.

29. Barbara Evans Clements, *Bolshevik Feminist: The Life of Aleksandra Kollontai* (Indiana University Press, 1979).

30. A. M. Blokh, *Sovetskii Soiuz v Inter'ere Nobelevskikh Premii* (Fizmatlit, 2005), pp. 144–152.

31. Ibid., pp. 201–216.

32. Ibid., p. 337. On Timofeev-Resovskii, see Raissa Berg, "In Defense of Timofeeff-Ressovsky," *Quarterly Review of Biology* 65 (1990), no. 4: 457–479; Daniil Granin, *Zubr* (Sovetskii Pisatel', 1987).

33. Blokh, *Sovetskii Soiuz v Inter'ere Nobelevskikh Premii*, pp. 272–278.

34. Ibid., pp. 333–334.

35. Ibid., p. 425.

36. Ibid., pp. 426–429.

37. Ibid., pp. 589–610.

38. Rajinder Singh and Falk Riess, "The 1930 Nobel Prize for Physics: A Close Decision?" *Notes and Records of the Royal Society of London* 55 (2001), no. 2: pp. 267–283. See also V. L. Ginzburg, *O Nauke, o Sebe, i o Drugikh*, third edition (Fizmatlit, 2004), pp. 408–411.

39. Ginzburg, *O Nauke*, pp. 412–413.

40. Michael Wines, "Nobel Winner Laments Poverty of Russian Science," *New York Times*, October 12, 2000.

41. Ibid.

42. "Vladimir Putin Nagradil Zhores Alferova ordenom 'Za zaslugi pered otechest-vom,'" at lenta.ru.

43. *Sankt-Peterburgskii Nauchno-Obrazovatel'nyi Tsentr RAN* (NOTs, 2007).

44. Interview with Alferov, St. Petersburg, January 13, 2009.

45. See http://www.school.ioffe.ru/school/. Also see *Litsei 'Fiziko-tekhnicheskaia Shkola'* (undated brochure, likely 2006).

46. http://www.school.ioffe.ru/readings/

47. Interview with Zhores Alferov, NOTs, St. Petersburg, January 19, 2006.

48. *Sankt-Peterburgskii Nauchno-Obrazovatel'nyi Tsentr RAN*.

49. Graham and Dezhina, *Science in the New Russia*, p. 10.

50. "Zhores Alferov potratit tret' Nobelevskoi premii na sozdanie fonda podderzhki obrazovaniia i nauki," at lenta.ru. See also Wines, "Nobel Winner Laments Poverty of Russian Science."

51. Richard Sakwa, *Putin: Russia's Choice* (Routledge, 2004); *Russian Politics Under Putin*, ed. Cameron Ross (Manchester University Press, 2004).

52. In December of 2008, masked and armed agents of the FSB raided Memorial, a non-government organization dedicated to the memory of those repressed in the Stalin era. Agents alleged that Memorial was connected with terrorist activities. In carrying out the raid, the FSB showed that it had much in common with Stalin's NKVD.

53. Gerald Easter, "The Russian State in the Time of Putin," *Post-Soviet Affairs* 24 (2008), no. 3: 199–230.

54. Paul R. Josephson, "Technological Utopianism in the Twenty-First Century: Russia's Nuclear Future," *History and Technology* 19 (2003), no. 3: 279–294.

55. Allison et al., *Avoiding Nuclear Anarchy*.

56. Rosatom is determined to modernize the nuclear energy industry, which had fallen on hard times since the Chernobyl disaster of 1986. Skeptics point out that the cost of building scores of reactors at billions of dollars each will waylay the ambitious program.

57. For example, Andrei Sakharov saw nuclear power as crucial to the future of civilization. He worried about limited reserves of fossil fuels and their contribution to greenhouse gases. Sakharov supported the development of "peaceful nuclear explosions"—small charges for use in excavation, creating underground caverns for storage of hazardous waste, closing runaway oil-well fires, and easing tension along fault lines to prevent earthquakes. He criticized the safety culture of the Soviet nuclear industry, but called for redoubled efforts to expand nuclear power production with the construction of nuclear reactors underground. To prevent more "Chernobyls," whether caused by human error, terrorists, earthquakes, or acts of war, Sakharov suggested an international law to forbid construction of reactors above the ground.

58. "Gosdum naznachila Zhoresa Alferova glavnym po iadernym otkhodam," at lenta.ru

59. Charles Digges, "Alferov to Head Nuclear-Waste Committee," *St. Petersburg Times*, July 13, 2001.

60. "Gosatomnadzor: Minoborony provotsiruet ekologicheskuiu katastrofu," at lenta.ru

61. "Akademika Alferova sniali s posta," at lenta.ru

62. http://www.aaas.org/news/releases/2006/1027nuclear.shtml

63. "Akademika Alferova sniali s posta," at lenta.ru

64. "Kommunist Zhores Alferov ushel s rukovodiashchego posta v Dume," at lenta.ru

65. Interview with Zhores Alferov, NOTs, St. Petersburg, October 24, 2007.

66. Ibid.

67. "Zhores Alferov: RAN vynuzhdena zanimat'sia buznesom a ne issledovaniiami," at lenta.ru

68. Galina Stolyarova, "Alfyorov Slams Science Funding Cuts," *St. Petersburg Times*, October 5, 2004.

69. Andrey Allakhverdov and Vladimir Pokrovsky, "Kremlin Brings Russian Academy of Sciences to Heel," *Science* 31 (2006), November 10: 917.

70. Iulia Taratuta, "Nobelevskii laureat uvolen s raboty," *Kommersant*, July 1, 2006.

71. Vladimir Ugriumov and Iurii Medvedev, "Chernye shary dlia akademika," at www.rg.ru

72. "Naglost' nado pobezhdat' doblest'iu." *Sovetskaia Rossiia*, August 5, 2006.

73. Ugriumov and Medvedev, "Chernye shary dlia akademika."

74. "Kak 'zakazal' Akademika Alferova? Andrei Fursenko i Iurii Kovalchuk spustia 15 Let Vziali Revansh za svoe Uvol'nenie iz FTI," at www.informacia.ru.

75. Arkadii Sosnov, "Konflikt v Fiztekhe kak zerkalo akademicheskoi reformy," at http://www.poisknews.ru. See also Iulia Taratuta, "Nobelevskii laureat uvolen s raboty," *Kommersant*, July 1, 2006. Zabrodskii acknowledged that the LFTI leadership ought to have asked Alferov privately to step down as scientific director rather than inform him of the need to give up the post at a council meeting. He thought it inappropriate that Alferov gave him the "director's watch," then precipitated "a personal conflict of interests." The conflict reflected Zabrodskii's concerns that Alferov's interests lay in NOTs and this would hurt the LFTI financially and programmatically. According to some observers, Zabrodskii took it personally that Alferov had been one of several academicians who spoke against his candidacy to full membership in the Academy during the last elections.

76. Sosnov, "Konflikt v Fiztekhe kak zerkalo akademicheskoi reformy"; "Kak 'zakazal' Akademika Alferova?"; "Naglost' nado pobezhdat' doblest'iu," *Sovetskaia Rossiia*, August 5, 2006.

77. Sosnov, "Konflikt v Fiztekhe kak zerkalo akademicheskoi reformy; "Kak 'zakazal' Akademika Alferova?"; "Naglost' nado pobezhdat' doblest'iu."

78. "Kak 'zakazal' Akademika Alferova?"

79. Fursenko, as cited at www.hitechno.ru/?page=news&eng=1.

80. http://www.anticompromat.ru/kovalch1/kovalch1bio.html

81. http://www.anticompromat.ru/putin/ozero.html

82. "Kak 'zakazal' Akademika Alferova?"

83. http://www.worldeconomics.ru/about/Xozdenieponaukamirina.html

84. http://www.worldeconomics.ru/about/Xozdenieponaukamirina.html

85. http://www.worldeconomics.ru/about/Xozdenieponaukamirina.html

86. Katherine Ters, "Nothing New Identified in Innovation, *St. Petersburg Times*, April 8, 2003.

87. "Kak 'zakazal' Akademika Alferova?"

88. For a good overview of industrial policy, see Richard D. Bingham, *Industrial Policy American Style: From Hamilton to HDTV* (M. E. Sharpe, 1998). See also Christian Ketels, "Industrial Policy in the United States," *Journal of Industry, Competition and Trade* 7 (2007), no. 3–4: 1566–1679. For skepticism about industrial policy, see Frederick Buttel, "How Epoch Making Are High Technologies? The Case of Biotechnology," *Sociological Forum* 4 (1989), no. 2: 247–261; Richard Brahm, "National Targeting

Policies, High-Technology Industries, and Excessive Competition," *Strategic Management Journal* 16 (1995), summer: 71–91.

89. For more on these issues, see Paul R. Josephson, "Russia's Nanotechnology Revolution," *Georgetown Journal of International Affairs* 10 (2009), no.1: 149–157.

90. Dan Medovnikov, Stanislav Rozmirovich, and Irik Imamutdinov, "Elephants on the Field," at http://eng.expert.ru.

91. http://www.thenanotechnologygroup.org/index.cfm?content=89

92. Alferov, "O programme Rossiiskoi Akademii Nauk v oblasti nanotekhnologii," *Vestnik Rossiiskoi Akademii Nauk* 78 (2008), no. 5: 427–435.

93. Andrey Allakhverdov and Vladimir Pokrovsky, "Russian Academy President Narrowly Wins Reelection," *Science* 320 (2008), June 6: 1270–1271. See also http://www.telegraph.co.uk/news/worldnews/europe/russia/2050057/Russian-scientists-reject-Vladimir-Putin-ally.html.

94. http://www.nanowerk.com/news/newsid=4994.php

95. Alferov, interview, St. Petersburg, January 13, 2009.

96. Taratuta, "Nobelevskii laureat uvolen s raboty."

97. "Energy Award Parts with Nobel Prize," based on Russian article published April 19, 2006, at www.kommersant.com. The source suggested that when Alferov took the award Putin was so outraged that he refused to attend the ceremony, though he was to present the prize.

98. Ugriumov and Medvedev, "Chernye shary dlia akademika."

99. *Pravda*, September 21–24 and September 25–26, 2007.

100. *Proceedings of the 10th International Symposium on Nanostructures: Physics and Technology* (St. Petersburg, June 2002), SPIE, volume 5023, ed. Zhores Alferov and Leo Esaki. For Alferov's paper, see pp. 357–364.

101. Zhores Alferov, stenogram, "Opening Speech (in Russian)," first meeting, fifth session, State Duma of Russian Federation, Moscow, December 24, 2007.

102. Ibid.

103. Ibid.

Index

Russian Revolution, 3, 4, 15–26, 33, 63,
102, 112, 198, 219, 228, 234, 242,
263
Rybakov, Anatolii, 74

Sagdeev, Roald, 211
Sakharov, Andrei, 5, 10, 11, 13, 34,
48–54, 102, 114–117, 135, 183–194,
200–210, 218, 224, 263
Semenov, Nikolai, 8, 103, 122, 231–233
Severodvinsk (Molotovsk), 8, 39, 60,
137, 138
Siberia, 15, 32, 33, 41–44, 72, 112, 173,
178, 212
Siemens, 12, 19, 172, 241
Skobeltsyn, Dmitrii, 101, 233
Smirnova, Liudmila Nikolaevna, 141
Smolnyi Institute, 112, 126
Socialist Realism, 24, 46–48, 91, 96,
115
Solzhenitsyn, Alexander, 96, 112
Soros, George, 213, 217, 226, 227
Sosunova, Maria Mikhailovna, 42, 43
Soviet Academy of Sciences, 5–12, 50,
53, 78, 96–101, 109, 119, 120, 124,
130–132, 136–140, 143–150, 157,
163, 172, 177, 181–183, 186, 190,
195, 198, 199, 203, 207–210, 222,
223, 232, 233, 247, 248, 255
Sozina, Nina Nikolaevna, 94, 106–108
Sputnik, 113–115, 118, 265
St. Petersburg, 1, 18–20, 24, 42, 86, 94,
200, 215, 218, 224, 247, 254
St. Petersburg Scientific Center, 12, 223,
248
Stafeev, V. I., 126, 131
Stalingrad, 15, 39–42, 67–72, 91, 241
Stalinism, 4, 7, 10, 24, 33, 34, 54, 57,
73, 78, 81, 92, 122, 124, 140, 177,
196, 229
Stalin, Joseph, 2, 5–8, 21, 26, 33, 39–45,
51, 56–59, 62–70, 73, 82, 90, 91,
95–106, 109, 111, 112, 115, 127, 138,

182, 185, 198, 203, 218, 232, 233,
255, 263, 265, 268
State Committee on Science and Technology, 130, 209, 210
Sverdlovsk, 60–63, 67, 70, 92, 121, 124,
219, 220
Szigeti, György, 142

Tamm, Igor, 5, 102, 194, 203, 231–233
Trotsky, Leon, 3, 21, 22, 28, 50, 100,
218
Tsimliansk, 90–94, 265
Tuchkevich, Vladimir, 8–10, 110, 119,
123–138, 143, 147–152, 160, 166,
189, 190, 202
Turinsk, 60–67, 73
Turkmenistan, 34, 203
Turksib Railway, 34

Ukraine, 22, 26, 38, 69, 78–80, 95, 109,
120, 124, 137, 203, 221, 222
Ukrainian Academy of Sciences, 72,
121
United Nations, 114, 221
United States, 3, 5, 9, 10, 18, 25,
47–49, 54, 89, 90, 97, 105, 110–118,
140–145, 157, 163, 169, 172–174,
178–181, 184, 185, 193, 200, 222,
241, 245, 255–258
University of Illinois, 142–145, 157,
185
Urals, 60–66, 72, 120
Uvarov, Anatolii, 126, 131

VAK, 101, 121, 122, 184, 185, 190–192
Valter, A. F., 146, 232
Vareikis, I. M., 23, 40
Varian Associates, 162, 165
Vavilov, Nikolai, 97, 98
Vavilov, Sergei, 97, 101, 120, 250
Velikov, Evgenii, 210, 211
Vitebsk, 6, 16, 17, 23, 26–31, 41
Volga River, 15, 78, 125